Knowledge-based Expert Systems in Chemistry
Not Counting on Computers

RSC Theoretical and Computational Chemistry Series

Series Editor:
Jonathan Hirst, *University of Nottingham, Nottingham, UK*

Titles in the Series:
1: Knowledge-based Expert Systems in Chemistry: Not Counting on Computers

How to obtain future titles on publication:
A standing order plan is available for this series. A standing order will bring delivery of each new volume immediately on publication.

For further information please contact:
Sales and Customer Care, Royal Society of Chemistry, Thomas Graham House, Science Park, Milton Road, Cambridge, CB4 0WF, UK
Telephone: +44 (0)1223 432360, Fax: +44 (0)1223 420247, Email: sales@rsc.org
Visit our website at http://www.rsc.org/Shop/Books/

Knowledge-based Expert Systems in Chemistry
Not Counting on Computers

Philip Judson
Consultant, Harrogate, UK

RSCPublishing

RSC Theoretical and Computational Chemistry Series No. 1

ISBN: 978-0-85404-160-2
ISSN: 2041-3181

A catalogue record for this book is available from the British Library

© Philip Judson 2009

Published by The Royal Society of Chemistry,
Thomas Graham House, Science Park, Milton Road,
Cambridge CB4 0WF, UK

Registered Charity Number 207890

For further information see our web site at www.rsc.org

Preface

Computers began to think, in their simple way, in the 1960s; they ceased to be mere adding machines. People interested in using computers to help with chemical synthesis design were among the earliest researchers in the field of artificial intelligence, and the results of their work have had a major impact on chemical software development ever since. I had the good fortune to become involved just at the time when the scientific community began to take chemical information and knowledge systems seriously – in the 1980s, twenty years after the pioneers had taken the lead. I have watched some of the systems grow from research ideas into mature products and in this book I write about them. The book is biased because I have written about what I know. However, I have been involved in some key areas. Knowledge-based and reasoning-based approaches are in routine use to predict or plan chemical reactions, to predict toxicity, and to predict metabolism and biodegradation, and spin-off from research into them has produced the best-known chemical structure and reaction database systems. There has not been a book specifically about them until now and I think it is time to fill the gap.

I came into this field almost by chance and, while chemistry has remained a central theme, it has taken me into biology, aspects of mathematics and theories of logic, and even psychology and law. I have crossed the boundaries between industry and academia and collaborated with scientists on every continent, and what I write is about the results of that collective effort. As everyone writes in the preface to a book, because it is true, it would be impractical to list the names of everyone to whom I owe gratitude. However, I do want to thank David A. Evans, A. Peter Johnson and Alan K. Long for inspiring my interest in knowledge-based systems in chemistry and their patient support while I developed an understanding of the science. I thank Alan K. Long, Anthony Long and Martin Ott for their comments and advice on some of the sections in this book.

RSC Theoretical and Computational Chemistry Series No. 1
Knowledge-based Expert Systems in Chemistry: Not Counting on Computers
By Philip Judson
© Philip Judson 2009
Published by the Royal Society of Chemistry, www.rsc.org

Contents

RSC Theoretical and Computational Chemistry Series No. 1
Knowledge-based Expert Systems in Chemistry: Not Counting on Computers
By Philip Judson
© Philip Judson 2009
Published by the Royal Society of Chemistry, www.rsc.org

CHAPTER 1

Artificial Intelligence – Making Use of Reasoning

Launched by half a dozen young men at a run, a three-metre long paper dart can fly successfully, dare we even say "gracefully", the length of a research station canteen before making an unfortunate landing in the director of research's Christmas lunch. It is just a question of getting the aerodynamics right. My school mathematics teacher reminded us on most days (several times on some) that all science is mathematics. But was it only the power of numbers he had in mind? Does science come down to a sweatshop full of equations mindlessly crunching numbers, real and imaginary?

Contrary to the perceptions of many people outside science, as well as too many inside it, science is not about proving facts: it is about testing hypotheses and theories; ultimately, it is about people and their opinions. Simple, rigid application of rules of aerodynamics may get you a paper dart that flies but in many fields human decision making is best supported by reasoned argument or the use of analogy and not much helped by numerical answers. The minimum braking distance for a car travelling at forty miles per hour is twenty-four metres, according to the Driving Manual from the Driving Standards Agency.[1] Assuming you can countenance the required mixing of miles and metres, does this information help you to drive more safely? Have you any more idea than I have how far ahead an imaginary twenty-four metre boundary-marker precedes you along the road?

And there is a further problem. "Numbers out" implies "numbers in", so what do you do if you have no numbers to put in? A regrettably popular solution is to invent them – or at least to come up with dubious estimates to feed into a model that demands them, which is close to invention. It is the only option if you want to apply numerical methods and to give numbers to the people asking for solutions. That numbers make people feel comfortable is a bigger problem than it may at first appear to be, too. Uncritical recipients of

RSC Theoretical and Computational Chemistry Series No. 1
Knowledge-based Expert Systems in Chemistry: Not Counting on Computers
By Philip Judson
© Philip Judson 2009
Published by the Royal Society of Chemistry, www.rsc.org

numerical answers tend to believe them, and to act on them, without probing very deeply. More sceptical recipients want to judge for themselves how meaningful the answers are but often find that the kind of supporting evidence associated with a numerical method is not much help. Many are the controversies over whether this or that numerical method is more precise but they are missing the point if the data are far less precise than the method. Perhaps numbers are unnecessary – even unsuitable – for expressing some kinds of scientific knowledge.

There are circumstances in which numerical methods are highly reliable. Aeroplanes stay up in the sky and make it safely to earth where they are supposed to do. Chemical plants run twenty-four hours a day, year in year out. Numerical methods work routinely in physical chemistry laboratories, and toxicology and pharmacology departments. But it is unlikely that the designers of the three-metre paper dart that took flight at the start of this chapter did any calculations at all. My guess is that they just went with a gut feeling based on years of experience making little ones.

This book is about uses of artificial intelligence (AI) and databases in computational chemistry and related science where qualitative output may be of more practical use than quantitative output. It touches on quantitative structure–activity relationships (QSAR) and how they can inform qualitative predictions, but it is not about QSAR. Neither is it a book about molecular modelling. Both subjects are well-covered in too many books to list comprehensively. A few examples are given in the references at the end of this chapter.[2-6] This book focuses on less widely described and yet, probably, more widely-used applications of AI in chemistry.

The term "artificial intelligence" carries with it notions of thinking computers but, as a radio personality in former times would have had it,[7] it all depends on what you mean by intelligence. If you type "Liebig Consender" into the Google™ search box, Google™ responds with "Did you mean *Liebig Condenser*" and provides a list of corresponding links without waiting for an answer. That is worryingly *like* intelligent behaviour whether it *is* intelligent behaviour or not. Arguments continue about whether tests for artificial intelligence such as the Turing test[8] are valid and whether a categorical test or set of tests can be devised. Perhaps it is sufficient to require that to be intelligent a system must be able to learn, be able to reason, be creative, and be able to explain itself persuasively. Currently, no artificial intelligence system can claim to have all of these characteristics. Individual systems typically have two or three.

To count as intelligent, solving problems needs to involve a degree of novel thinking, *i.e.* creativity. Restating the known, specific answer to a question requires only memory. Compare the following questions and answers. The first answer merely reproduces a single fact. Generating the second answer, simple though it is, requires reasoning and a degree of creativity.

"Where's the sugar?"
"In the sugar bowl".
"Where will the sugar be in this supermarket?"

"A lot of supermarkets put it near the tea and coffee, so it could be along the aisle labelled 'tea and coffee'. Alternatively, it might be in the aisle labelled 'baking'. Let's try 'baking' first – it is nearer".

One of the first computer systems to behave like an expert using a logical sequence of questions and answers to solve a problem was MYCIN,[9] a system to support medical diagnosis.

"Doctor, I keep getting these terrible headaches".

"Sorry to hear that. Is there any pattern to when the headaches occur?"

"Now you ask, they do seem to come mostly on Sunday mornings".

"And what do you do on Saturday evenings?"

The doctor's questions are not arbitrary. You can see how they are directed by the patient's responses. You can probably see where they are leading, too, but the doctor would still want to ask further questions to rule out all the possibilities before jumping to the obvious conclusion about the patient's Saturday nights out on the town. The aim of the MYCIN experiment was to design a computer system capable of choosing appropriate sequences of questions similarly, in order to reach a diagnosis efficiently.

This kind of reasoning is common throughout science although it often does not involve a dialogue; the questions may be implicit in a process of thought rather than consciously asked. Suppose you know that:

many α,β-unsaturated aldehydes cause skin sensitisation;

for activity to be expressed a compound must penetrate the skin;

compounds with low fat/water partition coefficients do not penetrate the skin easily:

many imines can be hydrolysed easily in living systems to generate aldehydes.

Actually, the story for skin sensitisers is better understood and can be more fully and more usefully described than this, but what we have will do for the purposes of illustration. Suppose you are shown the structure of a novel α,β-unsaturated imine and asked for an assessment of its potential to cause skin sensitisation. You will be aware that the imine might be converted into a potentially skin-sensitising aldehyde. If you have access to suitable methods you will get an estimate of the fat/water partition coefficient for the imine in order to make a judgement about whether it will penetrate the skin (most likely you will use a calculated $\log P$ value as a measure of fat/water partition coefficient, but there is more about that later in this book). You will presumably have the gumption to consider the partition coefficient for the aldehyde as well, in case the imine is unstable enough to hydrolyse on the surface of the skin.

Depending on the information, you will come up with conclusions and explanations such as:

"the query substance is likely to be a skin sensitiser because it has the right partition coefficient to penetrate the skin and the potential to be converted into an α,β-unsaturated aldehyde – a class of compounds including many skin sensitisers";

"the query is not likely to be a skin sensitiser because although it is an imine which could be converted into an α,β-unsaturated aldehyde – a class of compounds including many skin sensitisers – both compounds have such low fat/water partition coefficients that they are unlikely to penetrate the skin";

"the situation is equivocal because the imine has too high a fat/water partition coefficient to penetrate the skin easily but the related aldehyde has a lower fat/water partition coefficient and I do not know how readily the imine will hydrolyse to the aldehyde on the skin surface."

Systems in which a reasoning engine solves problems by applying rules from a knowledge base compiled by human experts were originally called "expert systems", on the grounds that they behave like experts. In this book they are distinguished by being called "knowledge-based systems". They use reasoning to varying degrees and they are creative in the sense that they solve novel problems and make predictions. The particular strength of the best of them is their ability to explain themselves. For example, there is fairly good understanding of why α,β-unsaturated aldehydes are skin sensitisers. The human compilers of a knowledge base can include that information so that the expert system can present it to a user when it makes a prediction and can explain how it reached its conclusion.

Given access to structures and biological data for lots of compounds, you might discover the rule that α,β-unsaturated aldehydes are often skin sensitisers, assuming you were not overwhelmed by the quantity of data. Knowledge-based systems as defined here make no attempt to discover rules from patterns in data – they simply apply the rules put into them by human experts. In terms of the criteria for intelligence, they are unable to learn for themselves. The more general term, "expert system", was later extended to include systems that generate their own models by statistical methods and apply them. While these systems are perhaps nearer to all-rounders in the stakes for showing intelligence than knowledge-based systems, they fall down on explaining themselves. They cannot go beyond presenting the statistical evidence for their rules.

A speaker remarked at a meeting I attended that "An expert system is one that gives the answers an expert would give . . . including the wrong ones". It might be fairer to compare consulting a knowledge-based system (which is what he was talking about at the time) with consulting a group of human experts rather than one, since knowledge bases are normally compiled from collective knowledge, not just individual knowledge, but his warning stands. Other people have, only half-jokingly, suggested that an expert system is one suitable only for use by an expert. That may be over-cautious but users of expert systems should at least be thinking and well-informed: it is what you would expect of someone taking advice from a team of experts.

References

1. N. Lynch and A. Wood, *The Driving Manual*, Her Majesty's Stationery Office, Norwich, England, 1992.
2. L. Eriksson, E. Johansson, N. Kettaneh-Wold and S. Wold, *Multi- and Megavariate Data Analysis*, Umetrics AB, Umeå, Sweden, 2001.
3. C. Hansch and A. Leo, *Exploring QSAR: Fundamentals and Applications in Chemistry and Biology*, Am. Chem. Soc., Washington DC, USA, 1995.
4. D. Livingstone, *Data Analysis for Chemists: Applications to QSAR and Chemical Product Design*, OUP, England, 1995.
5. T. Schlick, *Molecular Modelling and Simulations*, Springer, New York, 2002.
6. A. R. Leach, *Molecular Modelling: Principles and Applications*, Pearson Education EMA, Essex, England, 2nd edn, 2001.
7. C. E. M. Joad quoted by G. E. Penketh, *Anal. Proc.*, 1980, **17**, 163–4.
8. A. Turing, Computing Machinery and Intelligence, *Mind*, 1950, **50**, 433–60.
9. E. H. Shortcliffe, *Computer-Based Medical Consultation: MYCIN*, Elsevier Science Publications, New York, 1976.

Synthesis Planning by Computer

Organic synthesis chemists are used to working with ideas and rules of thumb. They are not inclined to plan reaction sequences to novel compounds on the basis of kinetic or thermodynamic calculations – indeed, they are rarely in the position to do so because data of sufficient reliability are not available for the calculations – but they have a reasonable success rate. How do they do it? Could a computer emulate the thinking of a chemist who works out a practical synthetic route to a complicated organic compound?

The tale is told of a conversation over a few beers one evening between three eminent chemists famed for their work in organic synthesis – Elias J. Corey, Alexander R. Todd and Robert B. Woodward. Corey, it is said, expressed the view that computers would eventually be capable of matching or even out-classing human reasoning; soon there would be machines capable of designing chemical syntheses just as well as chemists do. Todd and Woodward were sceptical, it is said, arguing that chemical synthesis was an art more than a science, calling for imagination and creativity well beyond the capacity of a computer. Corey saw how a computer might reason like a chemist and he proposed to set up a project to demonstrate the feasibility of his ideas. The story may be apocryphal but it does not matter if it is. The exciting thing is that Corey recognised a new challenge well beyond the everyday goals of most researchers and took it on. He was not alone in seeing and taking up the challenge – there were others who will feature in this chapter and the next – but his project proliferated like the mustard tree in the parable so that by now every chemist is familiar with at least one spin-off computer application that roosts in its branches.

Corey's project to develop a synthesis-planning program, OCSS ("Organic Chemical Simulation of Synthesis"), started in the 1960s and was described in a paper in *Science* in 1969.[1] By 1971, when a paper was submitted to the Journal of the American Chemical Society,[2] the program had been re-implemented as

RSC Theoretical and Computational Chemistry Series No. 1
Knowledge-based Expert Systems in Chemistry: Not Counting on Computers
By Philip Judson
© Philip Judson 2009
Published by the Royal Society of Chemistry, www.rsc.org

LHASA (Logic and Heuristics Applied to Synthetic Analysis) and the project was expanding.

Right from the start the plan was to develop a computer system that did not just think like a chemist, but communicated like one, too. Computer graphics was in its infancy. The computer mouse was yet to come to public notice – Douglas Engelbart filed his application for a patent in 1967[3] – but there were systems that linked a graphics tablet, or "bit pad", to a vector graphics screen (a line is displayed on a vector graphics screen by scanning the electron beam between the coordinates of the ends of the line, whereas in a television or a modern personal computer system the screen is scanned systematically from side to side and top to bottom and the beam is activated at the right moments to illuminate the pixels on the screen that lie on the line). Other researchers interested in using computers for chemistry were developing representations of chemical structures to suit computers, but in this project the computer would be expected to use the representations favoured by organic chemists – structural diagrams. In their paper in 1969,[1] Corey and Wipke wrote, "The following general requirements for the computer system were envisaged at the outset: (i) that it be an 'interactive system' allowing facile graphical communication of both input and output in a form most convenient and natural for the chemist . . . ".

A structural diagram is full of implicit information for a chemist that would not be perceived by someone not trained in chemistry. It is not a picture of a molecule, in as much as there can be a picture of one; it tells you what is connected to what, and how, but it does not tell you the three dimensional locations of atoms: like the map of the London Underground it is a graph. To make useful inferences, the computer needs to be able to "see" the graph like a chemist sees it, and so a chemical perception module in LHASA fills checklists for the atoms and bonds in a molecule for use in subsequent processing. For example, if a carbon atom is found to be bonded through a double bond to one oxygen atom and through a single bond to another oxygen atom which itself bears a hydrogen atom, the carbon atom can be flagged as the centre of a carboxylic acid group; if an atom is at a fusion point between two rings (which would have implications for its reactivity) it can be flagged as a "fusion atom".

Computer perception of a molecule may put the computer in the position to think about it the way a chemist would, but how does a chemist think of ways to synthesise even a simple molecule? The question embodies a host of others each of which probably has more than one answer. Corey would have been well-placed to look for answers suited to computer-implementation, having formulated his ideas for the retrosynthetic approach to chemical synthesis design for which he was later to receive a Nobel Prize in Chemistry,[4,5,6,7] and his thinking on the subject and his work on a computer system must surely have fed each other.

The essence of the retrosynthetic approach is that the target molecule contains the clues to the ways in which it might be constructed. That might be obvious but stating something explicitly and letting it lead your thinking can

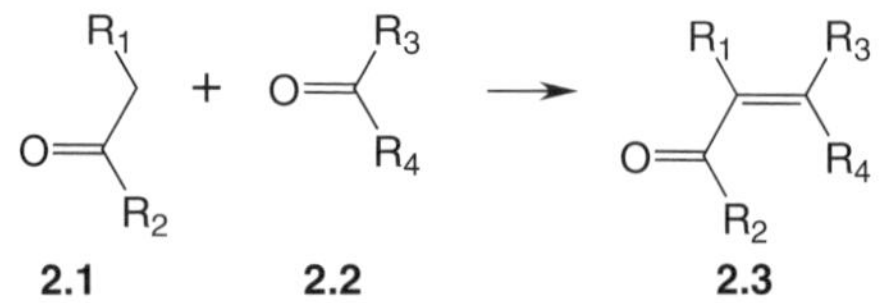

Scheme 2.1

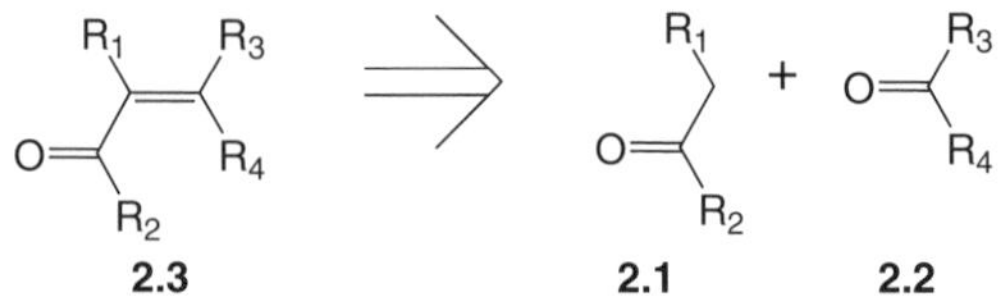

Scheme 2.2

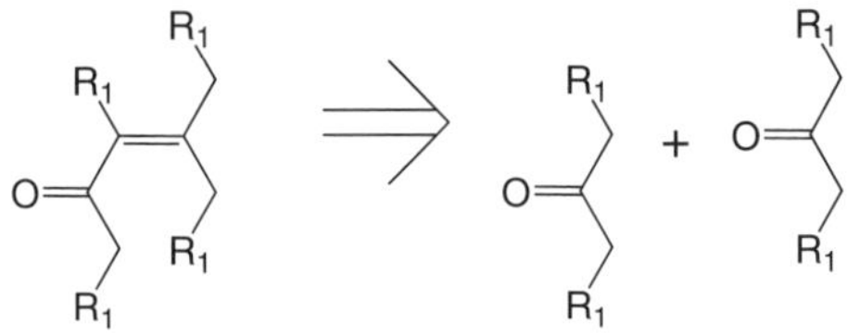

Scheme 2.3

completely change the way you tackle a problem. To take a simple example, the product of the aldol condensation is an α,β-unsaturated aldehyde or ketone (see Scheme 2.1). So if there is an α,β-unsaturated ketone in the target perhaps it could be made *via* the aldol condensation from the appropriate pair of ketones (or a ketone and aldehyde) as illustrated in Scheme 2.2. There is an obvious problem with this synthesis. The aldol condensation is likely to produce a mixture of products, only one of which will be the target, unless R^2, R^3, R^4, are all the same as $R^1.CH_2$ – in which case reactants 2.1 and 2.2 will be molecules of the same chemical and the retrosynthetic reaction can be written as shown in Scheme 2.3. To decide whether the aldol condensation is a good or bad choice for a synthesis, a chemist or a computer using the retrosynthetic approach needs access to a set of rules about the effects of appendages in the target represented by R groups in these schemes.

The LHASA knowledge base contains so-called "transforms" – descriptions of retrosynthetic reactions written in CHMTRN and PATRAN, languages which are described in Chapter 5. If the appropriate feature in a target structure is present it triggers the transform. Such triggering features are termed "retrons". The retron for the aldol condensation, for example, is a carbon-carbon double bond adjacent to a carbonyl group, sub-structure 2.4 in Scheme 2.4 in which the

Scheme 2.4

alpha substituents in positions left open can be hydrogen or carbon atoms. If the retron is present in a target structure the program reads the rules in the transform for the aldol condensation. There might be rules prohibiting the transform from proceeding if there are inappropriate heteroatoms or groups joined to the retron in the target structure. Other rules might increase or decrease the rating of the transform – a measure of how well it is likely to work – depending on the nature of substituents attached to the retron. If the rules allow the transform to proceed, the program uses instructions contained in the transform, the "transform mechanism", to generate the structures of the precursors from the description it has of the target.

Simply recognising functional groups in a target and applying the corresponding transforms to generate precursors is classed as an "opportunistic" approach in LHASA. It is an approach used frequently by chemists themselves to solve straightforward synthesis problems and certainly in the days when I sat my finals it underpinned many an examination question in organic synthesis . . . "Suggest practical ways of making six of the following structures". The eight or ten structures from which to make your choice would all contain features that the examiner hoped you would recognise as the clues to a neat synthesis. But for more challenging syntheses it has serious weaknesses.

If the target structure contains several retrons, which transform should you apply first? Do you have to try all the options? Even if, say, only three transforms are possible in the target structure and none of them fragment the structure, applying them in all possible orders gives you six synthetic routes to consider and fifteen reaction steps to assess in order to decide which route might work best (see Figure 2.1). If each of the transforms generates two precursors – as the aldol condensation transform does, for example – the numbers go up to twenty-four and thirty-nine, respectively. Actually, the situation is worse than that, because each time you apply a transform you generate new structural features that are themselves likely to be retrons for yet more possibilities.

With modern day computers a shot-gun approach to solving problems often works. You generate all possible solutions and then throw away the bad ones. But if you do the arithmetic for molecules of quite modest complexity you find there may be hundreds of thousands or even millions of potential solutions to a synthesis problem. The big issue when it comes to designing a chemical synthesis is not thinking of ways to do it – it is restricting your thinking to good ways of doing it.

The OCSS system included algorithms to prevent the inclusion of "useless or chemically naive structures" in the retrosynthetic tree (for example structures

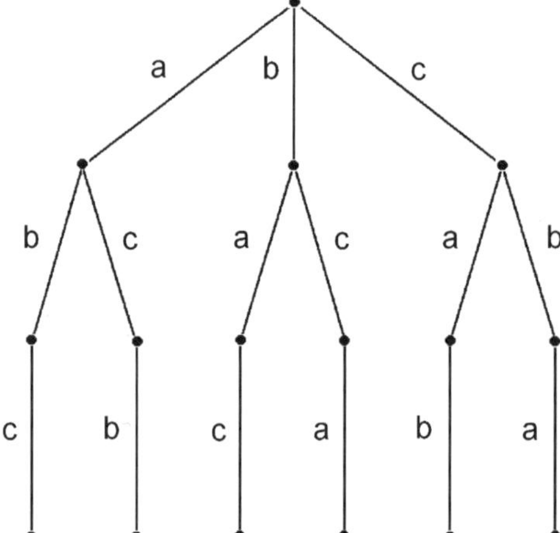

Figure 2.1 Representation of a retrosynthetic tree.

containing mutually incompatible functional groups) but the user was responsible for steering the analysis along productive lines. Understanding strategy in synthesis design and building it into the behaviour of the computer system was at the heart of the LHASA project. The aim was not to ignore the contribution that an experienced chemist using the system could make to an analysis, but to facilitate a dialogue between the user and the computer at a strategic level. For that to happen, the principles of different strategies needed to be built into the system.

Perhaps the most obvious strategic rule is one that says that a transform should only be applied if it leads to precursors that are simpler than the target, since there seems little benefit in choosing a precursor that is harder to make than the product. This may serve as the description of a strategy but it does not spell out what a chemist or a computer should be looking for. What amounts to simplification? An easy form of simplification to recognise is the fragmentation of the structure (retrosynthetically), since smaller molecules are usually easier to make, or to come by, than larger ones. There will be exceptions, of course. Perhaps the ideal precursor for your target is a readily-available natural product found as a glucuronide and to get what you want all you have to do is to saponify it. Surely, adding a glucuronide fragment to a structure (retrosynthetically) can hardly be described as a simplification? But examples like this will be oddities and do not deny the usefulness of the rule as a generalisation.

Other forms of simplification are a reduction in the number of bridged rings in a structure or a reduction in the number of stereocentres. In both cases, the reaction in the synthetic direction needs to be good at producing what is needed. There may be issues about the ordering of steps in a synthesis so that everything else about the molecule is optimum when the key reaction is applied.

So a strategy that just says "get rid of the bridges first, retrosynthetically" may not be sophisticated enough.

The idea of setting up a molecule to get maximum benefit from a particular reaction step leads to a different kind of strategy. If you study publications about syntheses of complicated structures certain reactions come to the fore as exceptionally useful. A reaction may be used in many contexts to control stereochemistry, another to create rings with similar substitution patterns in a variety of natural products. Concentrating on progressive retrosynthetic simplification is unlikely to reproduce these synthetic routes (or, more to the point, to come up with similar routes to novel structures). Alternative strategies are needed, based on an understanding of how to find the synthesis that makes most efficient use of the key step.

Turning aside from strategy for the moment, the problem of working backwards to a situation in which a particular retroreaction can be applied leads us to a tactic that proves to be important. Consider the synthesis of Structure 2.5 in Scheme 2.5.[8] Removing a side chain from a ring is generally a good move if you want to simplify a structure and the Friedel–Crafts reaction would allow just such a simplification by retrosynthetic removal of the acetyl group. The trouble is, the Friedel–Crafts reaction is not going to work for this molecule because the carboxylic acid group at the opposite end of ring does all the wrong things. It deactivates the ring, making it unlikely that the Friedel–Crafts reaction would proceed, but even if it did the substituent would end up in the wrong position because of the *meta*-directing influence of the carboxylic acid group. The trick is to convert the carboxylic acid group (retrosynthetically) into something activating and *ortho-, para*-directing – in this example the methyl group in Structure 2.6.

If LHASA operates step by step and is constrained to ignore non-simplifying transforms, it will never stumble upon the route in Scheme 2.5 because converting a carboxyl group into a methyl group really does not amount to a worthwhile simplification. One can think of many similar cases, requiring the conversion of one functional group into another or even requiring the introduction of a functional group as in Scheme 2.6,[8] adding to the complexity of the structure retrosynthetically. If a structure contains a very reactive or unstable functional group there is no sense in trying to do reactions in other parts of the structure. For example, a chemist would invariably remove an acyl halide retrosynthetically, converting it to a carboxylic acid, before considering any other chemistry and similar rules apply to

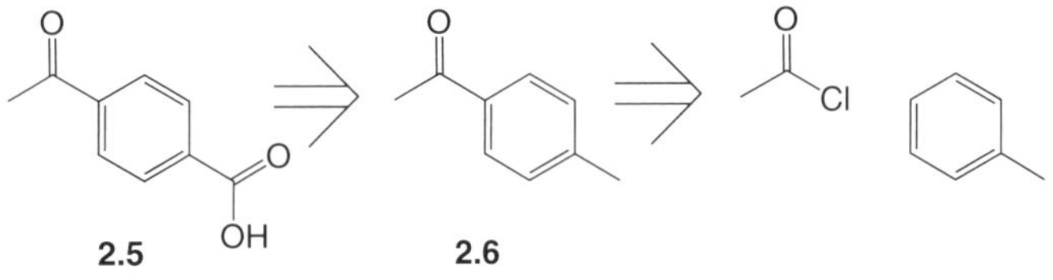

Scheme 2.5

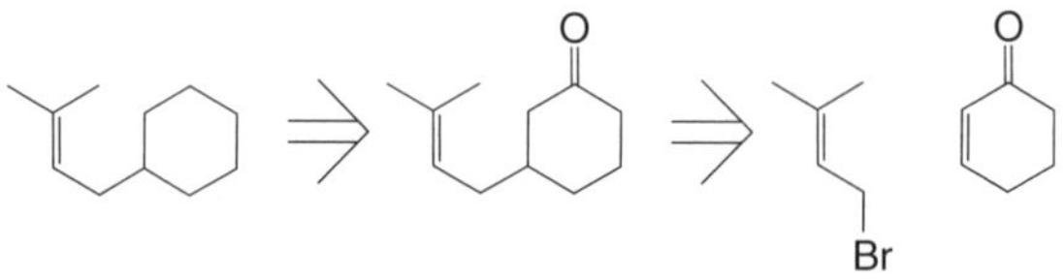

Scheme 2.6

several functional groups. A user might draw an enol as the target where it would make more sense to treat it as the related ketone or aldehyde, or *vice versa.*

The transforms in the LHASA knowledge base were classified mainly into goal and subgoal transforms, although later the scheme was modified so that a goal transform could also be called in the role of a subgoal for another one. The program will not apply a subgoal transform and present the result on the growing retrosynthetic tree unless it is the means to setting up a goal transform, *i.e.* a transform considered to be simplifying. In addition there are transforms labelled as "unmasking" to convert functional groups into simpler, related ones – for example to convert an imine into an aldehyde and an amine – and there are transforms that are automatically applied to convert presumed unstable structures into stable ones – for example to convert isolated enols into aldehydes or ketones. Even in the absence of strange targets that might be drawn by a user, the program needs to be able to recognise and deal with representations of functional groups in unstable forms because such a representation might be generated by the application of a transform mechanism to a structure containing unanticipated relationships between one functional group and another.

In 1985 Corey, Long, and Rubenstein described five strategies for synthesis planning in LHASA,[9] and Corey and Cheng presented them in more detail in their book ten years later.[5] In the LHASA program the strategies are presented in a menu prior to processing and the user chemist decides which one to apply to the current problem.

A *transform-based strategy* is one based on identifying a simplifying transform powerful enough to justify heavy use of subgoals so that it can be the key to the synthetic route. In rock climbing terms the key transform might be called the crux transform – if you can do that bit the hardest problem in the synthesis will be solved. They also use the term *long-range search* for this approach, since many retrosynthetic steps may need to be put in place before the goal transform can be applied. They list a score of reactions that are powerful enough to justify long-range searches, among them the Robinson Annulation and several versions of the Diels–Alder reaction.

A *structure-goal (S-goal) strategy* is one based on the identification of a potential starting material where a human would see a familiar pattern of atoms and bonds in the target. For example, a chemist looking at a target containing most of the features of the four-ring steroid system (Structure 2.7) would start by trying to find a way of making it from a naturally-occurring steroid in preference to constructing the ring system. The S-goal strategy, redesignated the *starting-material oriented strategy*, became one of particular interest to a group led by Peter Johnson collaborating on LHASA research in Leeds,[10,11] where associated developments led to spin-off described later in this book.

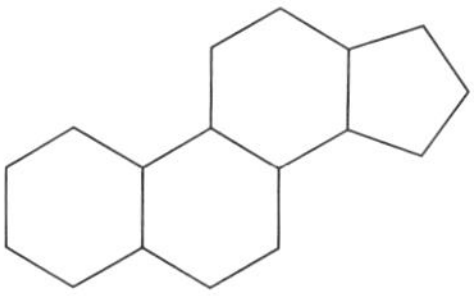

2.7

A *topological strategy* is one in which one or more bonds are identified whose disconnection leads to major molecular simplification. The usefulness of breaking the bond attaching a side chain to a ring has already been mentioned but there are more powerful applications of a topological strategy. Consider Structure 2.8 in Scheme 2.7. A reaction that made both the bonds marked *a* simultaneously would also create both rings in one step. But there is no reaction that can make those bonds and lead directly to Structure 2.8. Retro-synthetically, having identified the bonds as strategic it is necessary to do some sub-goal conversions to set up a goal transform. A Diels–Alder reaction might be a suitable goal, for example, as depicted in Scheme 2.7, depending on the nature of groups R^1 to R^4. Given Structure 2.8 as a target and asked to apply a topological strategy, LHASA would automatically find this retrosynthetic route among others.

A *stereochemical strategy* is one that gives priority to removing stereo-chemical centres (retrosynthetically) using transforms that allow efficient ste-reochemical control.

A *functional group oriented strategy* is one that recognises a pattern of inter-connected functionality in the target ideally set up for a simplifying transform (with or without modification of parts of the functionality through subgoals, such as converting an amide to an amine retrosynthetically). For example,

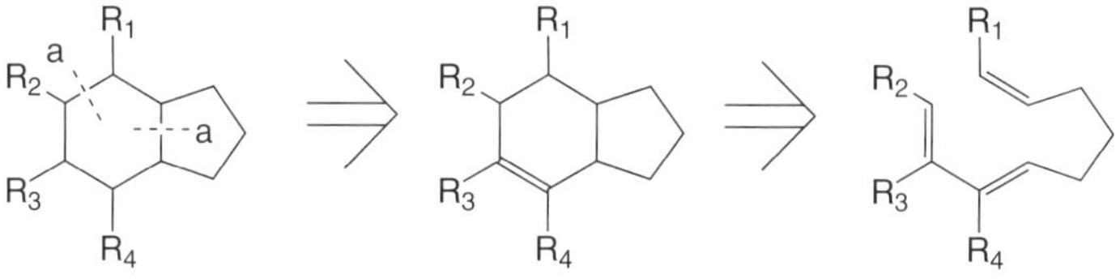

2.8

Scheme 2.7

groups R^1 to R^4 in Structure 2.8 might fortuitously be ideal for promoting the Diels–Alder reaction in Scheme 2.7, in which case the route would be favoured when using a functional group oriented strategy. In this case the choice is driven by the functionality of the target whereas with the topological strategy it was driven by the objective of breaking a particular pair of bonds.

LHASA continues to be available at the time of writing of this book,[12] now implemented in code that runs on Unix-based machines (including personal computers running Linux, for example) whereas it was originally in Fortran specific to Digital Equipment Corporation VAX computers which have become obsolete. New features have been added, such as a module called "LCOLI" to support work on combinatorial chemistry. But despite the amount of work put into its development, the ground-breaking achievements along the way, and the undoubted usefulness of the program in appropriate circumstances, LHASA does not seem to be much used. Likely reasons for this are discussed at the end of Chapter 3, but before that there are other projects and programs to write about.

References

1. E. J. Corey and W. Todd Wipke, Computer-Assisted Design of Complex Organic Synthesis, *Science*, 1969, **166**, 178–192.
2. E. J. Corey, W. T. Wipke, R. D., Cramer III and W. J. Howe, Computer-Assisted Synthetic Analysis. Facile Man-Machine Communication of Chemical Structure by Interactive Computer Graphics, *J. Amer. Chem. Soc.*, 1972, **94**, 421–430.
3. D. C. Engelbart, *X-Y Position Indicator for a Display System, US Patent*, 3 541 541, 1970.
4. S. Gronwitz, Presentation Speech for the Nobel Prize in Chemistry, 1990, in *Nobel Lectures, Chemistry 1981–1990*, ed. Bo. G. Malmström, World Scientific Publishing Company, Singapore, 1992.
5. E. J. Corey and X-M Cheng, *The Logic of Chemical Synthesis*, Wiley, New York, 1995.
6. E. J. Corey, Retrosynthetic Thinking – Essentials and Examples, *Chem. Soc. Rev.*, 1988, **17**, 111–133.
7. E. J. Corey, The Logic of Chemical Synthesis: Multistep Synthesis of Complex Carbogenic Molecules (Nobel Lecture), *Angew. Chem. Int. Ed. Eng.*, 1991, **30**(5), 455–465.
8. M. A. Ott, *Computer Methods in Synthetic Analysis. Applications to Reaction Retrieval and Synthesis Design*, Ph.D. thesis, Katholieke Universiteit Nijmegen, The Netherlands, 1996.
9. E. J. Corey, A. K. Long and S. D. Rubenstein, Computer-Assisted Analysis in Organic Synthesis, *Science*, 1985, **228**, 408–418.

10. A. P. Johnson, C. Marshall and P. N. Judson, Starting Material Oriented Retrosynthetic Analysis in the LHASA Program, *J. Chem. Inf. Comput. Sci.*, 1992, **32**, 411–417.

11. A. P. Johnson, C. Marshall and P. N. Judson, Some Recent Progress in the Development of the LHASA Computer Program for Synthesis Design, *Recl. Trav. Chim. Pays-Bas*, 1992, **111**, 310–316.

12. http://lhasa.harvard.edu/.

Other Programs to Support Chemical Synthesis Planning

For the most part this chapter is about programs that propose retrosynthetic routes for organic compounds using similar approaches, but this is also a convenient place to mention one or two programs that use different approaches and one that is for inorganic, rather than organic, chemistry. The chapter comprises a lot sections and skims over the details, but in defence of cataloguing so many projects I think there are things about each – sometimes technical, sometimes anecdotal – that will catch your interest, and in defence of not going into detail, if I get you interested enough to chase up the references I will have done a more useful job than paraphrasing what is eloquently put by the researchers themselves. Even this list of programs and projects is not exhaustive and you might like to seek out some of the publications I have missed, wherein might lie the makings of a future break-through. You might also enjoy an hour or two in a library with Volume 111 of Recueil des Travaux Chimiques des Pays-Bas[1] (which is in English) where you will find a whole series of papers about synthesis planning programs.

3.1 Programs that are Similar to LHASA in their Approach

3.1.1 SECS and PASCOP

Todd Wipke, who worked in E J Corey's team on OCSS and LHASA, went on to develop SECS (Simulation and Evaluation of Chemical Synthesis) which was more or less a sister program to LHASA, approaching the problem of synthesis

RSC Theoretical and Computational Chemistry Series No. 1
Knowledge-based Expert Systems in Chemistry: Not Counting on Computers
By Philip Judson
© Philip Judson 2009
Published by the Royal Society of Chemistry, www.rsc.org

design in much the same way.[2] SECS was adopted and adapted by a consortium of Swiss and German companies who used the name "CASP" (Computer Aided Synthesis Planning) for their project and product, and built a substantial knowledge base for it. The project ran for some years, at great cost, but was eventually abandoned – probably for the reasons suggested later in this chapter for the general decline of computer-assisted synthesis planning. An important innovation in CASP was the provision of a graphical knowledge base editor – although communication with end-users was graphical in LHASA and SECS, the transforms in their knowledge bases were described using typewritten codes (see Chapter 5).

SECS was also the basis for PASCOP, a program developed by Gérard Kaufmann and colleagues that concentrated on the synthesis of organophosphorus compounds.[3,4] The transforms for phosphorus chemistry in its knowledge base were later translated from the SECS/PASCOP knowledge base language ALCHEM into the LHASA knowledge base languages, PATRAN and CHMTRN, and incorporated into the LHASA knowledge base. By then I was working on the LHASA project in England, about which more is to come in Chapter 4. We were aware of the CASP and PASCOP groups and sought to collaborate with them. For business reasons and not through ill feeling it was soon clear that it was not practical to work with the CASP group. So it was with limited optimism that I arranged to visit Gérard Kaufmann in Strasbourg to talk about PASCOP, but he was a welcoming host and very open to collaboration. A simple deal was struck, without the aid of myriad lawyers that normally accompanies business negotiations, and we exchanged batches of new transforms about heterocyclic chemistry from the LHASA group for batches about phosphorus chemistry from the PASCOP group. In the long term, neither party got much payback from the exchange, since neither LHASA nor PASCOP went on to storm the markets, but lawyers may be relieved to learn that neither party lost out either.

3.1.2 SYNLMA

Peter Y. Johnson at the Illinois Institute of Technology (not to be confused with A. Peter Johnson at the University of Leeds) with colleagues at the Institute and at G. D. Searle and Company developed a synthesis planning program called SYNLMA[5]. They described the division of SYNLMA into three distinct modules, or, in their words, "independent units" – a chemical knowledge base, a user interface, and a reasoning component. They believed this structure to be unique. They were entitled to believe so and they certainly deserve to be acknowledged as the first to describe the architecture for a knowledge-based chemical synthesis planning system. LHASA, SECS, and some at least of the other chemical synthesis planning systems of the time, were similarly structured but none of the other research groups seem to have commented on its importance in their publications.

The separation of the reasoning component – commonly termed the "inference engine" or the "reasoning engine" – from the knowledge base is important. A simple, very general example illustrates why. Suppose you have an expert system that is driven by a set of knowledge rules that always take the form "If A is true then B is true". Later in this book we will come to rather more flexible expressions of relationships, but this will suffice for now. Supposing there is another rule saying "If B is true then C is true", given evidence that A is indeed true, it is the job of the reasoning engine to work out for itself that C must be true. The reasoning engine needs to know nothing about A, B, and C to work out the logical connection between the pair of statements, and if it can do the job in this case it can do it for any number of similar statements. It is not quite that simple, of course, because you must deal with contradictions and cycles (*e.g.* to cope with a third statement, "If C is true then A is true", without going into an infinite programming loop), but the point is that once the code is written it can work on any collection of "If . . . then . . . " statements without further modification. The knowledge base developer is free to document large numbers of rules and the reasoning engine soon outstrips the ability of the human brain to see all the implications of their interactions. The knowledge base developer does not need to write FORTRAN, Pascal, C++, Java, or whatever programming language is in vogue. The program can be designed to read simple textual statements, or an interface can be provided in which the user simply types terms like A, B, and C into boxes on a form.

SYNLMA as it stood in 1989 produced excessively large retrosynthetic trees – the so-called combinatorial explosion problem – and the authors described plans for increasing use of strategy to make the program more selective in its choice of which nodes on the tree to develop and which transforms to apply. However, it seems that the project terminated and SYNLMA disappeared from the scene.

3.1.3 SYNCHEM and SYNCHEM2

Herbert Gelernter and his colleagues opened their paper in Science in 1977[6] with a paragraph commenting on the changing attitude of chemists to computers. In the eight years since Corey and Wipke had first described OCSS, chemists – and scientists generally – had come to realise that computers were not restricted to number crunching but offered powerful ways to manipulate logic and reasoning. They mentioned that their interest in using artificial intelligence methods to solve chemical synthesis design problems at Stanford had started nine years before, leading to a first working version of SYNCHEM in 1971. So Corey and Wipke at Harvard may have been the first in the area to come to public notice, but the Stanford group had been with them on the starting grid.

Even if a search is limited to the most promising areas, doing a full retrosynthetic analysis takes a lot of computing time. With the power of current computers the waiting time for the final result would be unacceptable for real time operation; in the 1970s it would have been ridiculous. Corey's solution was to place the analysis under the interactive control of the user and to report the results step by step while processing continued in the background. Gelernter's group preferred the alternative solution of making the analysis a batch process. Solving a chemical synthesis problem does not require an immediate response and it would be acceptable to wait for a batch job if it had to run overnight, or even for a day or two.

The user entered structures into SYNCHEM in the form of Wiswesser Linear Notation (WLN) which was widely used in chemical information work at the time and is described in Chapter 5.1. It is a way of representing chemical structures uniquely and unambiguously in code that can be typed into a computer from a standard keyboard. It was thus ideal for a system intended for batch operation on the mainframe computers of the time.

The disadvantage with a batch process is that the user is not available to direct the progress of the analysis: SYNCHEM had to be able to solve all its own problems. So, for example, as a first step SYNCHEM had to choose what synthesis strategies it would try, whereas the user makes the choices for LHASA. Gelernter *et al.* term these initial options "synthemes", and they are based on functional groups and structural features in the query structure. Each syntheme has its own set of transforms, or, as Gelernter's group prefer to call them, schemas. As in LHASA transforms, each schema contains conditional statements driven by features in the query compound to raise or lower confidence in application of the retro-reaction represented in the schema, or to exclude it altogether in unfavourable cases. The merits of the resultant precursors are assessed and they are ranked. The highest-ranked one is chosen for further processing and the lower-ranked ones are returned to only if a subsequent step fails in the sequence that grows from the higher-ranked one. Processing stops when a successful retrosynthetic path has been found from the query structure to a structure in a catalogue of suitable raw materials held in a reaction library in SYNCHEM. At the time of writing of the paper, in 1977, the library contained the structures of about three thousand compounds from the Aldrich Chemical Company catalogue.

SYNCHEM2 was a replacement program, not just an upgrade of the existing one. The researchers felt that the SYNCHEM code had become too complicated for further adaptation and development, and they wanted to be able to deal with stereochemistry, which was not covered in SYNCHEM. It appears that SYNCHEM used WLN as an internal representation for structures as well as for the input interface, and this presented difficulties with structure and reaction manipulation. For SYNCHEM2 they used their own linear representation, SLING. A sentence in their paper explains why no-one should expect research projects that address worthwhile aims to deliver ready-made commercial products. It should be framed and displayed on the walls of board rooms and government grant agency offices:

"Building bridges into new terrain leaves little time or energy for filling potholes in the road behind".

3.1.4 SYNGEN

SYNGEN was also a batch program, developed by Hendrickson and colleagues at Brandeis University.[7,8] They placed particular emphasis on machine-applied strategy to generate reasonably-sized retrosynthetic trees containing the best routes. Their first step is to examine only the skeleton of the target structure and to try to break it into pieces that have the skeletons of known starting materials. Applied open-ended this analysis would generate too many candidates, of which many would be poor ones. For example, making a product from a dozen small components added one by one is unattractive compared with making it from just two larger starting materials that provide all the features that are needed. So first, they require that the chosen synthetic route be convergent.

Consider the retrosynthetic trees in Figures 3.1 and 3.2. Suppose that you want to make one kilo of 'a' and every reaction produces an 80% yield and that equimolar amounts of ingredients are used in all of them. In the convergent retrosynthesis in Figure 3.1 'a' is divided in half to give 'b' and 'c', and 'b' and 'c' are each similarly divided in half. In the non-convergent retrosynthesis in Figure 3.2 about one sixth of the target structure is chopped off at each step, ending up with 'f' and 'g' as the last two sixths. According to my arithmetic, you will need about 0.39 kg each of starting materials 'd' to 'g' if you follow the plan in Figure 3.1 – a total of just over 1.5 kg. If you follow the plan in Figure 3.2 you will need a total of about 1.7 kg of ingredients 'b″', 'c″', 'd″', 'e″',

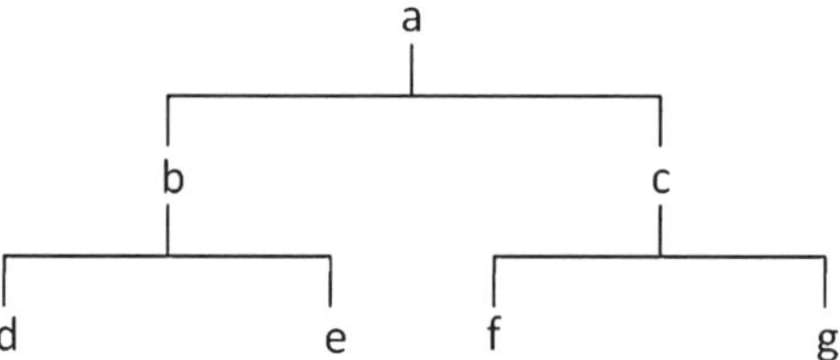

Figure 3.1   A convergent retrosynthetic tree.

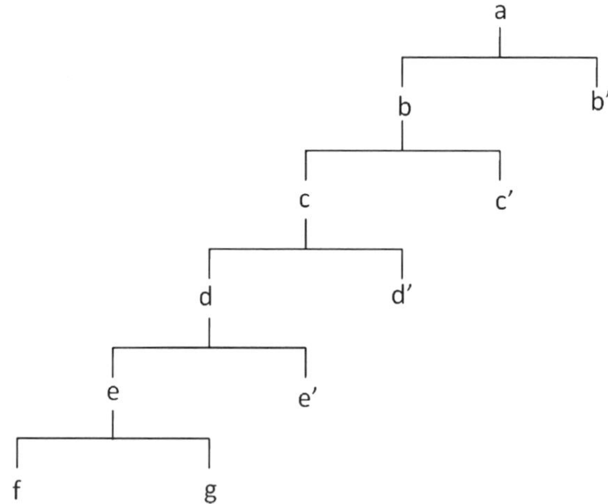

Figure 3.2   A non-convergent retrosynthetic tree.

'f', and 'g'. Not much of a difference and quite probably wrong anyway, bearing in mind that the calculation has just been done by the writer of a book about avoiding numbers. You might want to check the arithmetic for yourself and also to see what happens with more complicated trees or different percentage yields, but there is another consideration. The point about a convergent synthesis is not that structures should be divided into halves – that just made the arithmetic easier for the illustrations – it is that the tree is branched rather than linear. To follow the plan in Figure 3.1 you will have to do three reactions: to follow the plan in Figure 3.2 you will have to do five. That represents more time and effort, and each of the reactions is likely to involve similar amounts of waste in the form of disposable reagents, reaction solvents, heating, cooling water, and so on. Why carry out five reactions if you can get to the same end with three?

The designers of SYNGEN limited the size of the convergent tree by only allowing, as a first consideration, a maximum of two bonds to be cut in a retrosynthetic step and only allowing the process to go to two levels, making a total of four starting materials as in Figure 3.1. In another paper[9] Hendrickson and Huang described allowing the target to be divided into three fragments by a single step in cases where two of the resultant fragments are identical, since that is also a very efficient approach and one often applied in actual syntheses. They excluded futile fragmentations by requiring all four starting materials to have the skeletons of structures in the SYNGEN library of starting materials (what is the point of trying to fill in the details for a synthesis route if you cannot get the starting materials?). They limited the scope for creating large numbers of asymmetrical trees of progressively diminishing efficiency by not allowing the smaller of the two fragments created by splitting the target structure to be less than one quarter of the size of the target. By way of illustration of how effective this approach is, they quote the example of the estrone skeleton for which there are about 41 million possible assembly plans but only 1432 using the constraints they describe when used in conjunction with a catalogue of about 6000 starting materials.

Once a set of candidate retrosynthetic trees has been generated, the program seeks reactions from its knowledge base to bring about the required retrosynthetic disconnections. Reactions keyed by functionality already in the target are preferred over reactions requiring functional group addition or modification.

3.1.5 SYNSUP-MB and CAOSP

Malcolm Bersohn started work on the computer design of chemical syntheses at about the same time as Elias Corey and Herbert Gelernter. He described a program written in the LISP language to plan multi-step synthesis using a retrosynthetic approach in a paper that was published in 1972.[10] Each precursor as it is generated is given a score based on the degree of simplification from the target that is achieved and the "cost" of converting the precursor to the target compound through the forward reaction sequence that connects them. If the overall yield for a sequence falls below a predefined cut-off the precursor and the retro-reaction leading to it are automatically discarded. The "cost" is based on the expected overall percentage yield of the target from the precursor, the number of steps required, and the amount of material carried through the sequence. With regard to the amount of material carried through a sequence, Bersohn explains that a reaction with a 50% yield followed by one with a 100% yield is cheaper than a pair of reactions having yields in the reverse order – the overall yield is the same in both cases but in the first case less material that will ultimately go to waste is carried into the second reaction. The degree of simplification is measured in terms of the number of atoms and the number of functional groups added in going from the precursor to the target. Every time the program selects a structure for further processing from the retro-synthetic tree it chooses the one with the best score so far that does not already have precursors.

Bersohn and a group of Japanese colleagues have continued research and development of the program[11,12] leading to SYNSUP-MB at the Sumitomo Chemical Company and, according to Ryan Lilien, there may be further developments in collaboration with the CAOSP project.[15] There are publications describing the use of the program to design syntheses for real targets and there are references to a couple of examples of them at the end of this chapter.[13,14]

3.1.6 RESYN

Vismara *et al.* note that the overall approach to synthesis strategy is hierarchical and the design of Résyn reflects it.[16,17] At a first level, decisions are made about broad aims (*e.g.* whether a total synthesis is being sought or a partial one) and priorities (*e.g.* to develop a versatile plan that can be used for a set of related targets, to minimise the cost of the synthesis, *etc.*). The next level, the strategic level, is about deciding on the goals for dismantling the skeleton of the target retrosynthetically and the preferred order of doing them. The third level, the tactical level, is where retro-reactions are sought that can achieve the intended goals. Finally, at the fourth level you decide about the specifics such as reaction conditions and choice of protecting groups. This hierarchy is recognised by the designers of most of the other synthesis planning systems but it is not always described so clearly. The user can opt for an automated analysis or a step-by-step one under the user's direction, and Vismara *et al.* report that most users find the latter to be more useful.

Retro-reactions in Résyn are described in terms of the structural changes they bring about rather than in terms of retrons and how to modify them. A Résyn transformation is more generalised than a transform in LHASA – it is not tied to a retron, which in LHASA may itself be further limited by qualifying code (see Chapter 5.3). Structures and substructures are perceived by the program from three points of view: what Vismara *et al.* term "constitutional" (rings, chains, and how they are joined), stereochemical information (that is, just information about stereochemical centres and their inter-relationships), and functionality (locations and types of functional groups). The information, once generated is classified according to a hierarchical scheme to allow searching at varying levels from the specific to the general.

3.1.7 SOS, MARSEIL, CONAN HOLOWin and GRAAL

René Barone and Michel Chanon first published about their work on computer-aided chemical synthesis design in 1973 but they described a later version of SOS in 1990.[18] SOS generated all first level precursors to a target that it could do by applying the transforms in its knowledge base. It was for the user to decide which of these precursors were of interest for further processing, and so on, step by step. Interestingly, the primary purpose of SOS was originally for finding routes to heterocyclic compounds. Heterocyclic chemistry came much later in LHASA, the original priority having been to solve the problems of constructing skeletons such as the ones found in natural products that have long challenged synthesis chemists. Our own experience with LHASA when we introduced heterocyclic chemistry was also that the seemingly over-simplistic, unrestricted, "breadth first" approach used in SOS worked for simple heterocycles, whereas it was unsatisfactory for structures with difficult carbon skeletons.

In the program, MARSEIL, they introduced a graphical editor for input to the knowledge base (as distinct from input of query structures by an end user). This is a feature still lacking from most systems for synthesis planning. They provided for the user to enter his/her own perceptions about the current target during the course of an analysis and introduced self-learning into the program that took advantage of evaluations by users during analyses.

Over the years Barone, Chanon, and their colleagues have taken a different direction from most researchers in this area. Instead of concentrating on a machine to create retrosynthetic trees, they have turned their attention to providing chemists with tools to help them to design their own retrosyntheses. CONAN simply suggests the strategic bonds in a target to the chemist.[19,20] This is a much more approachable problem for a computer system than trying to work out an entire synthesis and it offers help with a task that humans can find difficult. The eye does not always see patterns of bonds in a complicated structure, especially if it contains fused and bridged rings of different sizes, whereas a computer can be depended upon to do so.

HOLOWin[21] seeks to apply key, simplifying transforms (the long-range search strategy in LHASA). The user chooses the transform of interest and asks HOLOWin to suggest how it could be used to make the target structure.

Instead of seeing it as a problem, Barone and colleagues make use of the capacity of a computer system to generate huge numbers of structures in GRAAL, which operates in the forward synthetic direction for situations where generating huge numbers is what is needed. For example, GRAAL suggests thousands of products from the thermal decomposition of thiamine,[22] Structure 3.1, which initially decomposes into about a dozen smaller molecules ranging from hydrogen sulphide to 1-methyl-3-amino-4-aminoethylpurine, which go on to react with each other and with subsequent products to generate several thousand in total. It would be both daunting and mind-numbingly tedious for a chemist to try to think up all the possibilities and write the structures on paper. On their website, Barone and colleagues suggest that GRAAL has many applications in aroma chemistry. I would add to that the potential for using it in food chemistry, if they do not already intend that to be covered by the term "aroma chemistry". I describe similar use of an adapted version of LHASA to predict the products of the Maillard process in Chapter 17.

3.1

You will find information about all of these programs on the Holowin website.[23]

3.1.8 AIPHOS and SOPHIA

AIPHOS[24] decides upon a strategy for the synthesis of a target structure using the topological and functional group based strategies that LHASA uses. Once the overall strategy for a synthesis has been decided upon, the program draws from a knowledge base of chemical reactions to construct a feasible implementation of it under the direction of the user. The user selects one of the strategic sites that the program has recommended (or makes his/her own choice of site) and the program generates possible precursors, without regard at this stage to the details of how the transformations might be brought about. The user selects a precursor, or set of precursors, and the program tries to complete the chemistry by introducing appropriate functionality. If a feasible path is found it is presented to the user.

A large database of available chemicals has been added to AIPHOS to support starting-material oriented searching.[25] The contents of the database are stored as graphs and reduced graphs at four different levels of detail (a reduced graph is one that describes connectivity but does not carry full information about the attributes of the nodes or arcs in it, as illustrated by the following scheme in AIPHOS): the first level consists of graphs that fully describe the starting materials; the next level contains information about positions and kinds of functional groups; the third level contains only information about the positions of the functional groups, *i.e.* without information about what the groups are; the fourth level contains only the skeletons of the structures. Having these different levels of detail in graph form allows the program to find the most complete fit that is available for a query quickly.

A sister program, SOPHIA, uses the same reaction knowledge base as AIPHOS to predict chemical reactions in the forward direction.[26]

3.1.9 Chiron

The Chiron program[27] seeks synthetic routes to stereochemically pure compounds by searching in a database of structures for maximum overlap of carbon skeleton, functionality, and stereochemistry. Alternatively, the user can specify a starting material which is to be mapped onto the target product. The program gives preference to starting materials that have the best overlap of functionality and stereochemistry with the target, but will allow functional group interconversions if necessary, taking into account retention or inversion of stereochemistry. The database includes non-chiral structures, such as commercially-available aromatic and heteroaromatic compounds, as well as chiral ones, to provide the program with a source of fragments that may be needed to complete the skeleton of a structure.

3.2   CICLOPS, EROS and WODCA – a Different Approach

EROS (Elaboration of Reactions for Organic Synthesis), growing out of CICLOPS, began as a system capable of modelling reactions in either direction – synthetically or retrosynthetically.[28] Later it was developed into two programs,[29] EROS for reaction prediction and WODCA for retrosynthetic analysis, and EROS has been expanded to support predictions about the metabolism of chemicals, reactions in a mass spectrometer, and for combinatorial chemistry.[30]

Johann Gasteiger and his group noted that in reality chemists do not try to solve synthesis problems by growing big synthetic or retrosynthetic trees in their imaginations: they look in the targets for substructures that remind them of available starting materials; in effect, they use a bi-directional search, growing backwards from the target and forwards from starting materials to find paths that connect. Gasteiger's group wanted to get away from the widespread synthon-based retrosynthetic approach and also from procedures for building synthesis trees that are limited to reactions that are already known. EROS and WODCA model reactions in a way that is more fundamental than the functional group approach, and they use properties such as bond polarity, inductive effects, resonance, and polarisability to evaluate the relative merits of breaking different bonds in the target (retrosynthetically).

The reactants and products in EROS, or the target, by-products and precursors in WODCA, are represented internally as bond and electron matrices (BE matrices). The rows and columns of the matrices refer to the atoms of a molecule or an ensemble of molecules (*e.g.* the set of reactants or the set of products for a reaction), the off-diagonal entries give the bond orders linking pairs of atoms, and the diagonal entries show the number of free electrons on an atom. Subtracting the BE matrix for the set of products of a reaction from the BE matrix for the set of reactants gives a new matrix that represents the reaction (an R matrix). Negative elements in the matrix indicate the breaking of bonds and positive elements the making of bonds. Changes to the distribution of free electrons on atoms appear in the diagonal entries of the R matrix. Rearranging the matrix equation for the purposes of reaction prediction the BE matrix for the products is the sum of the BE matrix for the reactants and

The BE matrix (left, reactant), the R matrix (middle) and the BE matrix (right, products) for the decomposition of formaldehyde cyanohydrin:

		1 O	2 C	3 H	4 H	5 H	6 C	7 N
1	O	4	1	0	0	1	0	0
2	C	1	0	1	1	0	1	0
3	H	0	1	0	0	0	0	0
4	H	0	1	0	0	0	0	0
5	H	1	0	0	0	0	0	0
6	C	0	1	0	0	0	0	3
7	N	0	0	0	0	0	3	2

+

		1 O	2 C	3 H	4 H	5 H	6 C	7 N
1	O	.	1	.	.	-1	.	.
2	C	1	.	.	.	.	-1	.
3	H	.	.	.	.	.	.	.
4	H	.	.	.	.	.	.	.
5	H	-1	.	.	.	.	1	.
6	C	.	-1	.	.	1	.	.
7	N	.	.	.	.	.	.	.

=

		1 O	2 C	3 H	4 H	5 H	6 C	7 N
1	O	4	2	0	0	0	0	0
2	C	2	0	1	1	0	0	0
3	H	0	1	0	0	0	0	0
4	H	0	1	0	0	0	0	0
5	H	0	0	0	0	0	1	0
6	C	0	0	0	0	1	0	3
7	N	0	0	0	0	0	3	2

Figure 3.3 Matrices representing the decomposition of formaldehyde cyanohydrins.

the R matrix, and for the purposes of describing retro-reactions subtraction of the R matrix from the BE matrix for the products gives the BE matrix for the reactants. Figure 3.3 illustrates this for the decomposition of formaldehyde cyanohydrin, the example used by Gasteiger and Jochum in their paper.[28]

Strictly, of course, in a real reaction the atoms labelled H^5 in both the starting material and the product in Figure 3.3 would rarely be the same atom – unlike, say, atom C^6, which really is the same atom on both sides of the equation. The mis-representation would become apparent if the reaction were carried out in the presence of deuterium oxide, since the model would fail to predict the deuteration you would expect as a direct consequence of the reaction mechanism. Ignoring the wanderings of protons keeps things simple and in practice it will not matter most of the time. What matters a lot is that you do not forgetfully drift into thinking that the model fully represents the true reaction mechanism. The topic of atom to atom mapping in reactions is mentioned again in Chapter 6.6.

Reactions normally involve the breaking of just one, two, or three bonds (converting a double bond to a single bond is also classed as a bond-breaking reaction in this context) and so it takes only a small set of R matrices to represent all the chemistry of potential interest in organic synthesis. The programs attempt to apply the transformations in all possible ways using the matrices for the candidate reaction participants, and rank the successful cases according to their favourability in terms of bond polarity *etc.* The first step is to identify breakable bonds in the reactants in the case of WODCA, or the target in the case of EROS – multiple bonds, bonds to heteroatoms, and bonds that are neighbours or next door neighbours of those features. Then the program can attempt to apply the reaction schemes, using the matrix arithmetic already described. However, there is a problem for EROS: given, say, ethyl acetate as a target, EROS would not be able to predict the usual method of synthesis from ethanol and ethanoic (acetic) acid because the product ensemble (and hence the product matrix) is incomplete; water is a product in addition to ethyl acetate.

Gasteiger and Jochum established that in the great majority of cases the by-product in reactions of this kind is one of a few, simple compounds such as water, carbon dioxide, nitrogen, hydrogen chloride, and sodium chloride. So EROS contains a small, standard set of simple synthesis partners to form an ensemble with target structures before the reaction schemes are applied.

The purpose of WODCA is not to generate synthesis solutions automatically, but to help a chemist with this bi-directional thinking. Different modules in WODCA offer help with the selection of potential starting materials based on mapping to the target structure, generating the structures of suitable synthesis precursors, and finding known synthesis reactions. The second of these, generating the structures of suitable synthesis precursors, is needed because it is often not possible to find mappings of starting materials directly if a target is structurally very different from the structures of available starting materials: it is better first to break the target down into potential intermediates. Users can activate the modules in WODCA in any order, and as often as they wish, as they explore the problem. WODCA links to the catalogues of several suppliers of fine chemicals and incorporates algorithms for tasks such as sub-structure and similarity searching.

WODCA is thus an example of a second generation of organic synthesis design programs which help a chemist to design a synthesis plan whereas earlier thinking was the other way round – that the computer should do the main job with help from the chemist.

3.3 PIRExS

PIRExS (Predicting Inorganic Reactivity Expert System) was developed by James P. Birk at Arizona State University.[31] Written in PROLOG, it predicts inorganic reactions, as the name implies, and is designed particularly with the teaching of inorganic chemistry in mind. The author sought to capture in the knowledge base the thinking of a group of experts on inorganic chemistry, pointing out that to build a database of all known and potential reactions would be a massive task and that even if it were done a user finding that a reaction was not predicted would not know whether that meant the reaction could not occur, was not yet reported to occur, or had simply been overlooked during compilation of the knowledge base – arguments that are relevant to knowledge-based systems in general if not always commented upon. He found that all the experts took the same approach. First, they categorised the reactants in the query into one or more classes and then they considered what reactions were appropriate to members of those classes. The program follows the same procedure.

Typical rules for reactant classes are ones like
A substance is an ionic compound if:
it contains two or more components;
one component is a metal;
another component is a non-metal;
the substance is uncharged.

To use PIRExS, a student enters one or two chemical formulae of inorganic compounds, ions, or elements and the program searches its knowledge base for a variety of types of reactions, including binary combinations of elements, redox reactions, disproportionation, displacement of hydrogen from water, steam, or acids, acid–base neutralisations, and many others. The student can ask to see the rules used to make the predictions and thus learn about the chemistry. PIRExS, an MS DOS application, has not been further developed recently, but at the time of writing of this book it is still available for download free to subscribers to the Journal of Chemical Education.[32]

3.4 COSYMA

COSYMA[33,34] uses information about sequences of reactions that have led to successful syntheses to propose strategies automatically for the synthesis of novel compounds. In this context, by "strategy" is meant the selection of a key sequence of chemical reaction steps, rather than "meta-knowledge". During a learning stage, COSYMA generalises known synthesis pathways, eliminating non-essential details about structures and compressing tactical steps along the road such as functional group interchanges and the attachment and removal of protecting groups. When presented with a target structure for which a synthetic route is required, the program constructs a generalised internal representation of the target and seeks suitable strategies from its library. Assuming that it finds some, they are applied to the target to generate putative synthetic routes. A final check is made for chemical validity before each solution is presented to the user, since incompatibilities with functional or topological features in the target may not have been evident in the generic representations used to match synthetic routes to the target.

3.5 CAMEO – Predicting Reactions

CAMEO[35], about which the first paper was published in 1980 in a series that continued for fifteen years,[36,37] is a system for predicting the products of organic reactions. It uses a knowledge base of information about reactions classed according to their type – *e.g.* base-catalysed reactions further sub-divided into elimination, addition, and substitution reactions. Rules about nucleophilicity, base strength, behaviour of different leaving groups, and steric accessibility are used to rank the reactions that a query structure might undergo; sections of the knowledge base dealing with pericyclic reactions take account of the Woodward–Hoffman rules;[38] and so on. CAMEO is complementary to the retro-synthesis planning systems. There are not many knowledge-based systems designed to predict forward chemistry, the others mentioned in this chapter being SOPHIA (Chapter 3.1.8) and EROS (Chapter 3.2).

3.6 What Happened to Synthesis Planning by Computer?

If you are an organic synthesis chemist do you use one of the knowledge-based computer systems to guide your planning? Whether organic synthesis is your speciality or not, did any of them feature in your undergraduate or post-graduate training? Unless you chose to study chemoinformatics at one of a tiny handful of universities I doubt if you have used any of them. Several synthesis planning programs are still around, but none have yet found widespread use. What became of computer-based synthesis planning?

In the mid 1980s Stuart Warren, who lectured to undergraduates on the retrosynthetic approach at the University of Cambridge, UK, ran a series of workshops for synthesis chemists at Chesterford Park Research Station near Saffron Walden. Each week small teams of chemists looked at synthesis problems that he brought along and then presented their solutions to the assembled workshop for scrutiny and discussion. At the end of each session a set of target structures was given to the chemists so that they could prepare their proposals for synthetic routes the following week. I was head of chemical information and computing at Chesterford Park and I went along to the workshops, collected the "homework" and invited LHASA to propose solutions.

Every week the chemists brought along good ideas for syntheses. Occasionally one proposal outshone the others but usually there was not a lot to choose between them. LHASA always came up with ideas, too. Often they were variants on what one or more of the chemists presented – for example, the same set of reactions but ordered differently in the route to the target – and occasionally they were identical. What LHASA turned out to be very good at – or perhaps it would be more correct to say the chemists were less good at than LHASA – was making use of rearrangements. Very frequently where LHASA used a rearrangement to solve a synthesis problem the chemists had missed it. But broadly speaking, LHASA was neither better nor worse than the chemists at proposing synthesis routes. Admittedly, LHASA was not really working alone – it was teamed up with a chemist (me) – but the chemists were working in collaboration, too.

It seems that LHASA did not fall out of favour because it failed to do what it was intended to do. Three things were to its disadvantage: instead of bringing fun to life it threatened to take fun away; it appeared to offer solutions to problems that did not need solving; not enough people used it often enough.

Designing a computer program capable of solving a newspaper crossword puzzle would present an interesting academic challenge and to succeed would be an achievement. Having created your crossword puzzle solver and implemented it on an easy-to-carry, pocket-sized computer, would you be able to sell it to commuters to use in the mornings as they took the train in from the suburbs? Probably not. Commuters like doing crossword puzzles. Chemists like designing syntheses. So why would they want to rid themselves of the task?

Actually, of course, if that was a problem, it was one of misconception. LHASA was never intended to take the job over: it was designed as an interactive system for a skilled chemist; a session with LHASA was intended to be like having a conversation with a group of other chemists over morning coffee. But most potential users of the system never got far enough with LHASA to find that out, because it failed to lure them into playing with it. Someone new to the program was confronted with learning about the LHASA project team's approach to strategies in synthesis before he/she could do anything useful. It may be true that thinking the way that LHASA thinks would make you a better chemist, but the program will not appeal to you if you have to make choices from option menus that you do not understand just to get started. Those of us who got to grips with LHASA know what you are missing – the program can be highly informative and motivating – but being told that is not going to sell it to you.

If LHASA offered solutions to problems that you could not otherwise solve, learning how to use it would be worth the trouble, but most of the time you can get by fine. As the workshops with Stuart Warren at Chesterford Park showed, chemists are good at solving synthesis problems without a lot of help. In any case, how often do you really have to make a particular chemical? Chemists in pharmaceutical and agrochemical research routinely have to make choices about which chemical to make out of a list of dozens, or hundreds, of candidates that are likely to have useful biological properties. Such chemists enjoy solving synthesis problems but their primary research interest is in what makes chemicals biologically active. If a chemical is hard to make, why choose it when easier candidates are equally suitable for the research in hand?

There remain situations in which a program like LHASA really would be useful: sometimes a difficult chemical does need to be made, to explore a theory about interaction with a biological site of action; a natural product may be of huge pharmaceutical interest, very hard to get from natural sources, and equally hard to synthesise: failure to think of a particularly neat way of synthesising a product may lead you to lose out to a competitor since having a better way to make something can be grounds for a new patent. But even though LHASA might save or earn a company millions of dollars it will be used only occasionally and this turns out to be a serious threat to the survival of a computer application.

I witnessed at first hand a battle between synthesis planning software and other software in which the synthesis planning software lost. It was LHASA that was vanquished, but it would have happened to any similar application. During a time of economic growth and high profitability for one company, research chemists brought in a variety of computer programs, including applications for synthesis planning, reaction databases, and chemical structure databases. When the economic climate took a downturn, the company looked at its computing budget and decided there had to be cuts. The financial department argued that its use of computers was essential to trading activities and to meeting legal obligations. Just about every department argued that word processing was a necessity. Gradually the options for economising were whittled down. What had started as a request to all departments to cut a few per cent from budgets quickly became an instruction to the research division to cut ten per cent. On receiving the instruction from the research director, each department insisted that its computing plans were immutable. After some fruitless meetings, the director asked for usage figures for all programs. In the chemistry department, the structure-based chemical information system holding information about in-house compounds was used on most days by most chemists – sometimes several times – and by many people in other departments, including biologists and toxicologists. The reaction database was used four or five hundred times per month. LHASA was used once or twice per month. So the director's edict went out: scrap LHASA – it is hardly used. Being rarely used is dangerous for the survival of a piece of software.

Finally, the people working on the development of synthesis planning software moved on. Computer-based synthesis planning came earlier than the development of structure and reaction database management systems. When researchers solving issues to do with computer processing of structures for synthesis planning began to think of simpler applications for their ideas – in structure and reaction database systems – and realised their commercial potential, some of them moved into those areas. Other groups came up with spin-off knowledge-based applications such as ones for predicting toxicity from structure, or predicting potential metabolic fate, which quickly became all-absorbing.

So chemical synthesis planning succumbed to the combination of the sheer difficulty of making headway with such a challenging problem, the rather infrequent need for such software among potential users and consequent limitations on funding, and the rapid development of exciting alternative opportunities for computer programming researchers.

Knowledge-based synthesis planning software may have fallen out of favour, but its pioneers made major contributions to the development of artificial intelligence and chemical information systems that are now central to chemical research, and they fundamentally changed the way research chemists think and organic chemistry is taught. There are few parts of the rest of this book that did not grow directly out of chemical synthesis planning research, and probably none that owe nothing to it. And being currently out of favour does not always spell the end. Out of the programs mentioned in this and the previous chapter some are still available and being developed, including LHASA, WODCA, and the programs from the teams led by Barone and Chanon. Researchers in Japan have reported successful practical use of SYNSUP within the last couple of years[13,14] and, intriguingly, according to Ryan Lilien's website exciting secrets will soon be revealed about developments coming out of collaboration with the SYNSUP project.[15]

References

1. *Recl. Trav. Chim. Pays-Bas*, 1992, **111**.
2. W. T. Wipke, G. I. Ouchi and S. Krishnan, Simulation and Evaluation of Chemical Synthesis - SECS: an Application of Artificial Intelligence Techniques, *Artif. Intell.*, 1978, **11**, 173–193.

3. C. Laurenco, L. Villien and G. Kaufmann, Synthèse Assistée par Ordinateur de la Phosphacarnegine-I. Etablissement du Plan de Synthèse avec l'Aide de Pascop, *Tetrahedron*, 1984, **40**, 2721–2729.

4. C. Laurenco, L. Villien and G. Kaufmann, Experimentation du plan de synthèse établi avec l'aide de pascop : Synthèse Assistée par Ordinateur de la Phosphacarnegine-II, *Tetrahedron*, 1984, **40**, 2731–2740.

5. P. Y. Johnson, I. Bernstein, J. Crary, M. Evans and T. Wang, Designing an Expert System for Organic Synthesis, in *Expert Systems Applications in Chemistry*, ed. B. A. Holme and H. Pierce, ACS Symposium Series, Am. Chem. Soc., Washington DC, 1989.

6. H. L. Gelernter, A. F. Sanders, D. L. Larsen, K. K. Agarwal, R. H. Boivie, G. A. Spritzer and J. E. Searleman, Empirical Explorations of SYNCHEM, *Science*, 1977, **197**, 1041–1049.

7. J. B. Hendrickson and A. G. Toczko, SYNGEN Program for Synthesis Design: Basic Computing Techniques, *J. Chem. Inf. Comput. Sci.*, 1989, **29**, 137–145.

8. J. B. Hendrickson, Systematic Synthesis Design. 6. Yield Analysis and Convergency, *J. Am. Chem. Soc.*, 1977, **99**, 5439–5450.

9. J. B. Hendrickson and P. Huang, Multiple Constructions in Synthesis Design, *J. Chem. Inf. Comput. Sci.*, 1989, **29**, 145–151.

10. M. Bersohn, Automatic Problem Solving Applied to Synthetic Chemistry, *Bull. Chem. Soc. Jpn.*, 1972, **45**, 1897–1903.

11. M. Takahashi, I. Dogane, M. Yoshida, H. Yamachika, T. Takabatake and M. Bersohn, The Performance of a Noninteractive Synthesis Program, *J. Chem. Inf. Comput. Sci.*, 1990, **30**, 436–441.

12. T. Takabatake, T. Kawai, A. Tanaka, W. Katoda, M. Bersohn and D. Gruner, *Joho Kogaku Toronkai Koen Yoshishu*, 2004, 27, 23–24 (Japanese).

13. A. Tanaka, T. Kawai, T. Takabatake, N. Oka, H. Okamoto and M. Bersohn, Synthesis of an Azaspirane via Birch Reduction Alkylation Prompted by Suggestions from a Computer Program, *Tetrahedron Lett.*, 2006, **47**, 6733–6737.

14. A. Tanaka, T. Kawai, T. Takabatake, N. Oka, H. Okamoto and M. Bersohn, Finding Synthetically Versatile and Common Intermediates for Multiple Useful Products with the Aid of a Synthesis Design System, *Tetrahedron*, 2007, **63**, 10226–10236.

15. http://www.cs.toronto.edu/~lilien/lilien_projCSP.html.

16. P. Vismara, P. Jambaud, C. Laurenço and J. Quinqueton, RESYN: Objets, Classification et Raisonnement Distribué en Chimie Organique. Langages et Modèles à Objets: Etats et Perspectives de la Recherche, *Collect. Didactique INRIA*, 1998, **19**, 397–419.

17. P. Vismara, J. Regin, J. Quinqueton, M. Py, C. Laurenço and L. Lapied, RESYN: Un Systeme d'Aide à la Conception de Plans de Synthèse en Chimie Organique, in *Actes des 12èmes Journées Internationales sur les Systèmes Experts et leurs Applications AVIGNON '92*, 1992, **1**, 305–318.

18. P. Azario, M. Arbelot, A. Baldy, R. Meyer, R. Barone and M. Chanon, Microcomputer Assisted Retrosynthesis (MARS), *New. J. Chem.*, 1990, **14**, 951–956.

19. F. Barberis, R. Barone, M. Arbelot, A. Baldy and M. Chanon, CONAN (CONnectivity ANalysis): a Simple Approach in the Field of Computer-Aided Organic Synthesis. Example of the Taxane Framework, *J. Chem. Inf. Comput. Sci.*, 1995, **35**, 467–471.

20. R. Barone and M. Chanon, Search for Strategies by Computer: the CONAN Approach. Application to Steroid and Taxane Frameworks, *Tetrahedron*, 2005, **61**, 8916–8923.

21. R. Barone, F. Barberis and M. Chanon, Le Logiciel HOLOWin, une Approche Simple, Rapide, Originale pour Rechercher des Stratégies de Synthèse par Ordinateur, *Sci. Chim. Lett. Dép. Sci. CNRS*, 1997, **69**, 14–16.

22. R. M. Barone, M. C. Chanon, G. A. Vernin and C. Parkanyi, in Generation of Potentially New Flavoring Structures from Thiamine by a New Combinatorial Chemistry Program, in *Food Flavor and Chemistry; Explorations into the 21st Century*, ed. A. M. Spanier, F. Shahidi, T. H. Parliment, C. Mussinan and E. Tratras Contis, Royal Soc. Chem., Cambridge, 2005, pp. 175–212.

23. http://www.holowin.u-3mrs.fr/.

24. K. Funatsu and S. Sasaki, Computer-Assisted Organic Synthesis Design and Reaction Prediction System, *"AIPHOS"*, *Tetrahedron Comput. Methodol.*, 1988, **1**, 27–37.

25. K. Satoh, S. Azuma, H. Satoh and K. Funatsu, Development of a Program for Construction of a Starting Material Library for AIPHOS, *J. Chem. Software*, 1997, **4**, 101–107.

26. H. Satoh and K. Funatsu, SOPHIA, a Knowledge Base-Guided Reaction Prediction System – Utilisation of a Knowledge Base Derived from a Reaction Database, *J. Chem. Inf. Comput. Sci.*, 1995, **35**, 34–44.

27. S. Hanessian, J. Franco and B. Larouche, The Pschobiological Basis of Heuristic Synthesis Planning – Man, Machine and the Chiron Approach, *Pure Appl. Chem.*, 1990, **62**, 1887–1910.

28. J. Gasteiger and C. Jochum, EROS – a Computer Program for Generating Sequences of Reactions, *Topics. Current. Chem.*, 1978, **74**, 93–126.

29. J. Gasteiger, The Prediction of Chemical Reactions, in *Chemoinformatics – a Text Book*, ed. J. Gasteiger and T. Engel, Springer-Verlag, Heidelberg, 1990, pp. 542–567.

30. R. Höllering, J. Gasteiger, L. Steinhauer, K. Schulz and A. Herwig, Simulation of Organic Reactions: from the Degradation of Chemicals to Combinatorial Synthesis, *J. Chem. Inf. Comput. Sci.*, 2000, **40**, 482–494.

31. J. P. Birk, Predicting Inorganic Reactions, in *Expert Systems Applications in Chemistry*, ed. B. A. Holme and H. Pierce, ACS Symposium Series, Am. Chem. Soc., Washington DC, 1989.

32. http://jchemed.chem.wisc.edu/JCESoft/Issues/Series_B/3B1/prog1-3B1.html.

33. P. Jauffret, C. Ostermann and G. Kaufmann, Using the COSYMA System for the Discovery of Synthesis Strategies by Analogy, *Eur. J. Org. Chem.*, 2003, 1983–1992.
34. L. Ellermann, P. Jauffret, C. Ostermann and G. Kaufman, Evolution of the Concept of Synthesis Strategy in the COSYMA System: Introduction of the Synthesis Variant, *Liebigs Ann. Rec.*, 1997, 1401–1406.
35. W. L. Jorgensen, E. R. Laird, A. J. Gushurst, J. M. Fleischer, S. A. Gothe, H. E. Helson, G. D. Paderes and S. Sinclair, CAMEO: A Program for the Logical Prediction of the Products of Organic Reactions, *Pure Appl. Chem.*, 1990, **62**, 1921–1932.
36. T. D. Salatin and W. L. Jorgensen, Computer-Assisted Mechanistic Evaluation of Organic Reactions. 1. Overview, *J. Org. Chem.*, 1980, **45**, 2043–2051.
37. J. M. Fleischer, A. J. Gushurst and W. L. Jorgensen, Computer Assisted Mechanistic Evaluation of Organic Reactions. 26. Diastereoselective Additions: Cram's Rule, *J. Org. Chem.*, 1995, **60**, 490–498.
38. R. B. Woodward and R. Hoffmann, in *The Conservation of Orbital Symmetry*, Academic Press, New York, 1970.

CHAPTER 4

International Repercussions of the Harvard LHASA Project

The paper aeroplane that took off in the first sentence of this book really did fly. Apart from providing a useful illustration of successful technology done without numbers, and having achieved the rather more pressing objective from my point of view of getting you to buy the book, it has one small further contribution to make. The place where it flew will feature from time to time in the pages that follow, and the plane and its flight in some way symbolise the pioneering spirit that was there.

Chesterford Park Research Station, on a hill near Saffron Walden in England, may or may not have been out of the ordinary for its time when I worked there in the 1960s and '70s, but it seemed like a novelist's evocation of an earlier time. Social interactions had a quaint formality, with senior staff addressed as Dr This and Dr That or by their initials, but known by nicknames when they were out of earshot. Scientific symposia took place in "the Mansion", where delegates sat in wicker arm chairs in an oak-panelled room with a beautifully-decorated ceiling and large windows looking out across the croquet lawn to an ornamental lake and trees. But on one day each year, the day of the Christmas lunch, there was uproar culminating in the lunch itself. The air was filled with screwed-up wads of paper, paper darts, and even bread rolls, as departments engaged in battle across the tables. Party streamers marked out paths like tracer bullets over the heads of an infantry armed with water pistols.

The paper aeroplane was a stunt dreamed up by the biologists. Their team delayed entry until the battle was in full cry before bursting in through the main doors of the canteen and launching their ultimate weapon. It was a ridiculous prank but it was also a grand gesture. It happened in a place where being brave and believing you could break new ground were virtues. Chesterford Park championed creativity, not market constraints and cost–benefit analysis. They came into play in later stage projects but not at the leading edge of research.

RSC Theoretical and Computational Chemistry Series No. 1
Knowledge-based Expert Systems in Chemistry: Not Counting on Computers
By Philip Judson
© Philip Judson 2009
Published by the Royal Society of Chemistry, www.rsc.org

The head of chemistry to whom I reported when I first went to work there, Geoffrey Tattershall Newbold – "GTN" in memos and nicknamed "Tatters" because of his second name – told me the job of a head of department was to maintain an environment in which chemists were free to invent.

When Geoffrey Newbold retired he was replaced by David Evans who brought new ideas and sharpened enthusiasm. One of his colleagues from university days was A. Peter Johnson, recently returned from working with E. J. Corey on the LHASA project at Harvard, and David Evans invited him to give a talk about it at Chesterford Park. What Peter described was hugely exciting. The IBM PC had barely been invented and not hit the market. Most computers were expensive "mainframe" machines and we did not even have a computer at Chesterford Park. The biologists sent their data on punched cards for processing in Felixstowe. And yet Corey, Johnson, and others were drawing chemical structures on computer screens and getting proposals for synthetic routes displayed. I cornered David Evans to ask how we could get involved. He wanted to set up a collaboration with Peter Johnson and I was offered the job of liaison man at Chesterford Park.

Peter Johnson built up a small team of researchers in Leeds to work on aspects of the LHASA project, as well as continuing laboratory research into organic synthesis. The team worked in day-to-day co-operation with the Harvard team in Cambridge, Massachusetts, using computer-to-computer communication which was not to get noticed by the public and named "email" until several years later. One of my colleagues from Chesterford Park, Graham Rowson, and I became part-time members of the team, travelling to Leeds about once a month to work through weekends when we could get computer time.

The Leeds LHASA group concentrated on the development and use of a pattern representation of transforms for the LHASA knowledge base. Before that, transforms were described in terms of sixty-four predefined functional groups. It had had the advantage of allowing the information to be processed using sixty-four bit numbers but the disadvantage of restricting LHASA to the chemistry of those groups. Two kinds of transform were recognised: "one group" and "two group", keyed respectively by the presence of one functional group or two in the target molecule. For example, the Grignard reaction of a ketone with a bromoalkane is represented by a one group transform in the knowledge base, since, as illustrated in Figure 4.1, it is characterised by a single functional group in the product. The aldol reaction is represented by a two group transform, being characterised by two groups in the product – a ketone and a double bond on adjacent carbon atoms. The groups in a two group transform can be more widely separated, an example being the ring opening of

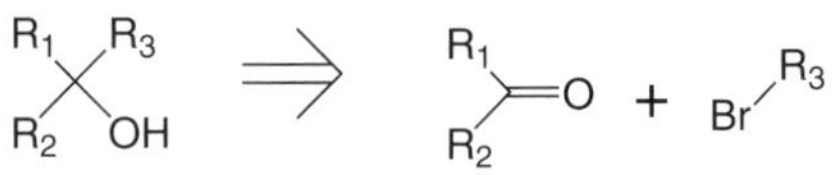

Figure 4.1   Transform for the Grignard reaction – a single group transform.

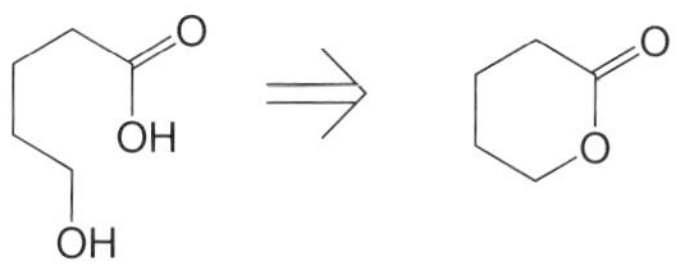

Figure 4.2 Transform for lactone ring opening – a two group transform.

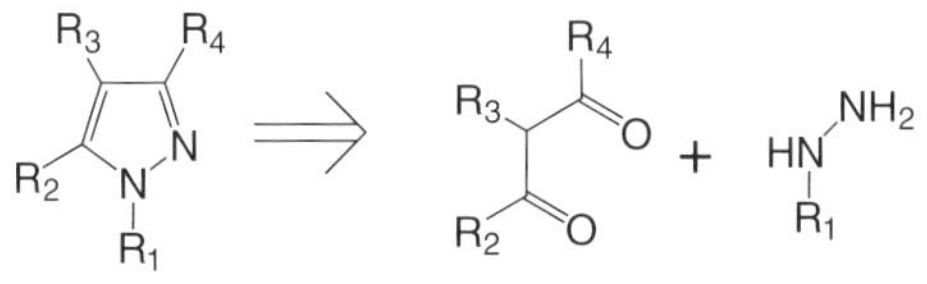

Figure 4.3 Transform for formation of a pyrazole – a pattern transform.

a gamma lactone in Figure 4.2, and the transforms are classified in LHASA according to the path length between the two groups.

The Leeds team worked on the development of PATRAN,[1] a language that allowed knowledge base writers to describe pretty well any kind of sub-structural pattern and use it to key a transform. For example the keying pyrazole substructure for the transform in Figure 4.3 can be represented in PATRAN simply as N%N%C%C%C%@1. The original one and two group transforms in LHASA were not rewritten, except for a few that were restricted too much by being tied to the specific functional groups that had been used to key them, but it was now possible to add huge amounts of chemistry that could not have been covered by the functional group approach – carbocyclic and heterocyclic aromatic chemistry in particular, which caught the attention of the fine chemical, pharmaceutical, and agrochemical industries. A collaborative users' group was set up as a forum for organisations in the UK interested in the project.

It soon became apparent that there were two issues to be addressed if LHASA was to become commercially useful, and that a collaboration between industry and academia was the way to tackle them. The program itself needed further development involving academic research of the kind suited to the groups at Harvard and Leeds. Having said that the program was already rather clever – to make an analogy with a human, it just did not know much. Most of the intellectual content of LHASA was to do with strategy, concepts, and so on. The knowledge base was adequate to support and demonstrate the program's "thinking" capabilities but was far short of describing the world of chemistry. The program understood about aromaticity and tautomerism, it could apply chemists' rules of thumb such as Markownikov's rules, or the rules about *ortho-*, *para-* and *meta-*directing groups in aromatic substitution chemistry, but its knowledge base did not contain anywhere near enough transforms. For example, it still knew hardly anything about aromatic heterocyclic chemistry. You do not get a PhD for entering existing knowledge into a computer and it would not be fair to expect an academic group to take on the job. In any case, there would be limitations on how well a young academic researcher working at the interface between computing and chemistry could assess information about synthetic heterocyclic chemistry. On the other hand, the very members of the users' group who wanted the LHASA knowledge base to be expanded were experts with access to relevant, specialised knowledge about synthetic reactions within their own organisations.

Each organisation knew more than the others about its particular areas of interest, but would they be willing to pool the knowledge? ICI Fine Chemicals Division at Blackley near Manchester were sponsoring the work in Leeds and Peter Bamfield, their representative, came forward with the idea of seeking support from the Department of Trade and Industry (DTI) to set up a more formal organisation. In the USA, E. J. Corey had attracted only limited support for a similar scheme but it was hoped that the greater acceptance of collaborative schemes in UK business culture would make a difference. Peter Bamfield, Peter Johnson, and I put together proposals for the formation of a not-for-profit company limited by guarantee, assisted by the company secretary for Fisons Agrochemicals Division, the owners at that time of Chesterford Park. In essence, members would make commitments both to sponsor the work through modest financial contributions and, more importantly, to donate knowledge by writing new entries for the knowledge base. In order to get both financial and practical support and to make the scheme available to companies of different sizes and research capacity, two kinds of membership were proposed: full members would pay a modest membership fee and make donations of knowledge in kind and in return would have use of LHASA with the expanded knowledge base on-line free of charge and, if they had a licence from Harvard for the program, they would be able to have the collaborative knowledge base on site and to keep it in perpetuity; associate members would pay the membership fee but would be allowed access to the program and knowledge base only on-line and on a pay-as-you-go basis (although the term "pay-as-you-go" had not come into vogue then).

The proposals were accepted by the members of the informal LHASA Users' Group that already existed and an application for funding to the DTI was granted. Thus LHASA UK Limited was born in 1983, with the primary objects of sponsoring research work at Harvard and Leeds on the LHASA program and expanding its knowledge base. There had been scepticism about the idea of knowledge sharing between competing companies, with concerns either that companies would refuse to donate or that they would select worthless material to donate, but the scheme worked. Even knowledge from public sources donated by full members had proprietary value because it carried with it the expert interpretation provided by specialists working for the members, and there were frequently supporting snippets of knowledge coming from unpublished experiments at the members' sites. Senior management at one site noticed that a member of staff was working close to half time on researching the public literature to support the LHASA project while easily-available internal knowledge that was not commercially sensitive was being passed over. It was deemed more cost-effective to donate the proprietary knowledge to LHASA and that began to happen.

Nine UK-based companies signed the memorandum and articles of the company at its inauguration, but within months a first non-UK European company had joined and within a year a USA-based company had joined. The name "LHASA UK" had been chosen to make it distinct from the LHASA group at Harvard, but the terminal "UK" was not clearly understood. So, by agreement with Harvard, "UK" was dropped. The company became, and remains, Lhasa Limited.[3] At the height of interest in the LHASA program more than twenty big companies were users of the program, many of them having copies in house.

Lhasa Limited has grown into a major player in the promotion and use of knowledge-based expert systems, especially in the fields of toxicology and metabolism, and its story will continue in succeeding chapters, but it was not the only important development to come out of the Harvard project.

SECS and PASCOP have already been mentioned in Chapter 3. SECS, in particular, led on to greater things. Todd Wipke and his colleagues saw the potential for a chemical structure database management system. The result was MACCS and they set up a company to market and develop it, Molecular Design Limited. Within a few years MACCS had swept the market worldwide; almost every company with a significant interest in research in organic chemistry was using it, and "MDL" was a household name within the chemical community. MDL soon developed a sister product to MACCS, REACCS – a chemical reaction database management system which was unwittingly to contribute to a decline in interest in LHASA as users realised that chemical reaction searching was really what they were looking for, once they saw it in action, rather than synthesis planning. Over the years the company and its products have evolved. MACCS was superseded eventually by ISIS Host and ISIS Base and surely every chemist has used ISIS Draw at some time – the tool developed by MDL for graphical input to its applications and for general chemical drawing work. After a series of mergers and name changes the company is now part of Symyx.

Peter Johnson also realised that many people used LHASA as though it were a reaction searching tool, albeit not designed for the purpose, and he developed a reaction database management system, ORAC – actually a bit ahead of the development of REACCS. He set up an eponymous company to market it, and a chemical structure database management system, OSAC, soon followed.

One of E. J. Corey's PhD students, Stew Rubenstein, had one of the tiny, early Apple Macintosh computers to play with and spent his evenings (and according to the reports I have heard, large parts of the nights) developing the world's first chemical drawing package some time before ISIS Draw was thought of – indeed, before Microsoft Windows had become established. It was ChemDraw,[4] another program that just about every chemist has seen or used. It made his fortune and launched the company, CambridgeSoft.

Somewhere along the line, MDL was acquired by Maxwell Communications Corporation (MCC) and so was ORAC Limited. Robert Maxwell was well-known for his newspaper interests but his empire included major scientific publishing houses, his career having started with bringing chemical literature out of Berlin at the end of the war. MCC looked after the electronic side of publishing. Maxwell did not merge MDL and ORAC but preferred to let them compete, apparently in order to keep them fighting fit.

Being part of MCC, ORAC was drawn into Scitechinform[2] – what might have been the most significant venture to hit the chemical information world in half a century. Maxwell, seeing opportunity in the changes that were looming in Russia, planned to launch a direct competitor to CAS, the Chemical Abstracts Service of the American Chemical Society. He brought together an advisory council of experts from Russia, East Germany, Bulgaria, West Germany, Switzerland, the United Kingdom, and the USA. He intended that the new product should be fully electronic from day one, and ORAC's role was to provide the technology. We planned to base it on HTSS, a fast structure-searching system from Hungary. I was working for ORAC by then and represented the company at the inaugural meeting of the council at the Waldorf Astoria hotel in New York. It was an exciting meeting but an exhausting one – not least because I was there to demonstrate the software on an IBM PC and had never even met DOS until a couple of days earlier, all my computing having been on DEC VAX computers.

A few months later Robert Maxwell fell off the back of his yacht in the Mediterranean. He was a big man and the waves rolled round the world. Scitechinform was no more. MDL went through a troubled period but survived. ORAC was closed down and its core staff promptly formed Synopsys, out of which came the chemical database add-ons, Accord for Excel and Accord for Oracle. Synopsys later became part of Accelrys, adding two more company names and product names to the list of those whose lineage traces back to E. J. Corey's LHASA project.

References

1. G. Hopkinson. *Computer-Assisted Organic Synthesis Design*, PhD thesis, University of Leeds, England, 1985.
2. "Soviet Deal for Maxwell", *New York Times*, Reuters, 7th April 1989.
3. Lhasa Limited, 22–23 Blenheim Terrace, Woodhouse Lane, Leeds, LS2 9HD, UK. http://www.lhasalimited.org.
4. ChemDraw is supplied by CambridgeSoft Corporation, Cambridge, Massachusetts, USA. http://www.cambridgesoft.com.

CHAPTER 5

Structure Representation

Although computer graphics came early in the development of applications for organic chemistry, there was a significant period, beginning well before the start of the synthesis planning projects and continuing for some years after it, during which graphical input of structures was not practical for most people. The majority of computers were so-called mainframes and users communicated with them *via* punched cards, punched tape or, if they were lucky, directly from a line printer keyboard. Ways were needed of encoding chemical structures in a form that could be entered as text. IUPAC nomenclature was too lengthy and chemical names, being so complicated, were not suitable for direct processing in a computer program and did not lend themselves to easy conversion into representations that would be. William Wiswesser, among others, was already formulating ideas for representations based on minimal strings of characters in the late 1940s, but widespread interest and activity came to the fore with the founding of the Chemical Notation Association (later to become the Chemical Structure Association) in 1965 in the USA and a UK division of it in 1969.

5.1 Wiswesser Line-Formula Notation

Researchers were looking for a code to represent structures that could easily be stored and searched as well being the means for entering structures into computer programs and retrieving them. Several schemes were proposed but the one that became popular for a while was Wiswesser Line-Formula Notation (WLFN, or WLN).[1-4] In WLN, carbon atoms in unbranched chains are represented by integers, the integer being the number of carbon atoms in the chain, and structural features such as branching points, unsaturation, and hetero atoms are represented by letters. Hetero atoms with single letter chemical symbols, such as O, N, and S, are so represented in WLN. Two-letter

RSC Theoretical and Computational Chemistry Series No. 1
Knowledge-based Expert Systems in Chemistry: Not Counting on Computers
By Philip Judson
© Philip Judson 2009
Published by the Royal Society of Chemistry, www.rsc.org

chemical symbols are capitalised and placed between hyphens (*e.g.* Fe becomes -FE-) – many computer systems at the time when WLN was developed only recognised upper case letters. A few elements and functional groups that are found very frequently in organic compounds have their own symbols in WLN. Chlorine, for example, is "G", and an -NH$_2$ group is "Z".

There are complicated rules about the order in which a structure must be described, to ensure that there is a unique WLN, of which the most fundamental is that the correct notation is the one that comes latest in an alphanumerically ordered list. So, for example, given that Q is the code for an hydroxyl group, and simple carbon chains are represented by the number of atoms they contain, ethanol might be 2Q or Q2. Q2 is the correct code, since it comes later in an ordered list than 2Q. The rules get very complicated when it comes to dealing with polycyclic and bridged ring systems, but WLN was in use and under progressive development for long enough for most eventualities to have been provided for. Figure 5.1 contains some examples of structures and their WLN codes taken from a paper by Graham Palmer.[3] You may be able to work out how some of the simple ones are encoded, but look upon the encoding of the tricyclic structure as a significant challenge. If you can get most of the way to understanding it without referring to the papers referenced at the end of this chapter you should be working in code-breaking for a government somewhere, not sitting reading books about chemical information.

A cunning use of WLN in the days when most people had no access to a computer but wanted to be able to do sub-structure searches was the permuted index. In an ordinary ordered list, words, phrases, or codes appear once each, in alphabetical order, the primary sorting being on the first character in the code, and the entries in the list are left justified. In a permuted index, each code

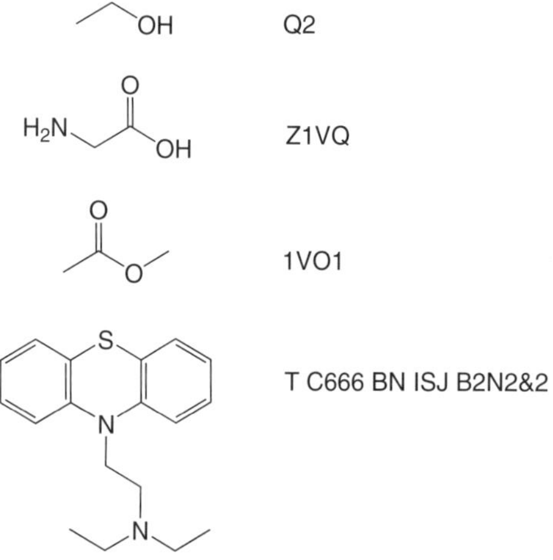

Figure 5.1   Some examples of Wiswesser Line Notations from the paper by Graham Palmer.[3]

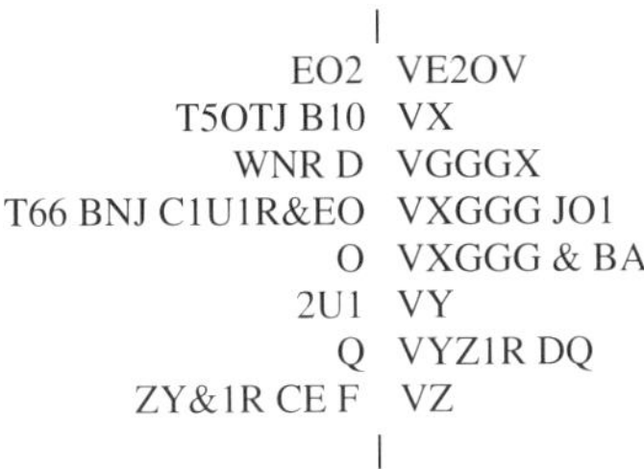

Figure 5.2 A section in a permuted index of Wiswesser Line Notations from the paper by Graham Palmer.[3]

appears repeatedly in the list, ordered by each letter in the code in turn. To make it easy to search the list by eye, the letter currently being used for ordering is aligned down the centre of the page. Figure 5.2 is a simple illustration based on an extract from the paper by Graham Palmer.[3] It shows a small part of the section under "V" in a permuted index of WLN. Suppose you are looking for structures containing a trichloroacetyl group ($CO.CCl_3$). In WLN code a trichloroacetyl group will be either "VXGGG" or "GXGGV" depending on how the encoding rules operated for the structure. Compounds containing "VXGGG" are grouped together in this part of the index. To complete the search you would need also to refer to the section of the index where "GXGGV" was listed under "G". This method of searching is limited compared with present day sub-structure searching, but in its time it was a good deal better than having no sub-structure searching.

5.2   SMILES, SMARTS and SMIRKS

Dave Weininger, working with Corwen Hansch and Al Leo, devised a much simpler code for representing chemical structures with a character string.[5,6] It is the widely adopted SMILES (Simplified Molecular Input Line Entry System). The user is not expected to apply rules in order to ensure there is only one way of encoding a structure – indeed, it is usually possible to work out many valid SMILES codes for the same structure – the computer generates the same canonical SMILES code regardless of how the user enters it[7] (see Chapter 6.1.1).

The rules for writing SMILES codes can be downloaded from the web site of Daylight Chemical Information Systems Inc.[8] In principle, they are very simple. Choose whatever atom you wish in the structure and write down its atomic symbol within square brackets. Choose one of its neighbouring atoms. Write down a symbol to represent the order of the bond: "–", "=", "#", or ":" for single, double, triple, or aromatic bond respectively followed by the atomic symbol for the neighbouring atom in square brackets. If there is more than one neighbouring atom, use parentheses to show the branching. Separate the parts of disconnected structures (*e.g.* the anion and cation in a salt) by a period mark.

The square brackets can be omitted for the elements commonly found in organic structures – B, C, N, O, P, S, F, Cl, Br and I – as long as they are in their minimum normal oxidation state in the structure. For these elements, when represented without square brackets, aromaticity can be implied by using lower case instead of upper case letters. The symbols for single and aromatic bonds can be omitted: the bond order will be assumed to be single unless the attached atoms are written in lower case, as just described, to make them aromatic.

Hydrogen atoms can be included explicitly as atom types or omitted, subject to rules which follow. Implicit hydrogen atoms are automatically assumed to make up for unsatisfied valencies of elements that are written without square brackets so that, for example, "CC" is automatically recognised as ethane, $CH_3.CH_3$ (the distinction between implicit and explicit hydrogen atoms is discussed in Chapter 7.2). The hydrogen count for atoms within square brackets must be specified also within the square brackets. Charges on atoms are shown within the square brackets.

If there is a ring in a structure, imagine breaking a convenient bond in the ring and omit that connection from the SMILES string. Label the two atoms with a single digit number to show that they are connected. For example, a SMILES code for cyclohexane is C1CCCCC1 and a code for benzene is c1ccccc1. It would be a strange choice to make, but C(CC1)CCC1, for example, would also be a valid code for cyclohexane.

The SMILES system provides for representation of stereochemistry and it allows you to specify isotopes and not just elemental types for atoms. Figure 5.3 shows some structures with SMILES codes taken from the theory manual on

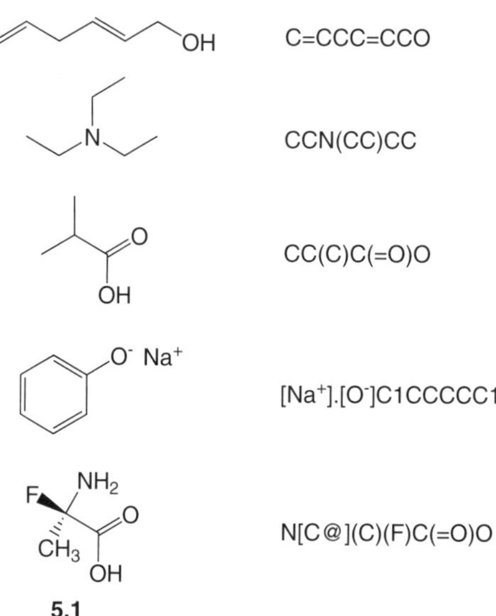

Figure 5.3   Some examples of SMILES codes from the Daylight Software website.

the website of Daylight Chemical Information Systems Inc.[8] One of the examples illustrates how tetrahedral stereochemistry is represented. Imagine looking down the bond towards the stereocentre in Structure 5.1 from the atom attached to it that is first listed in the SMILES string – the nitrogen atom. If the other atoms taken in the order listed in the SMILES code are distributed anticlockwise as you view them (which is the case in this illustration), represent the stereochemistry with the "@" symbol. This symbol was chosen because "@" provides a reminder in itself, having a "tail" that runs anticlockwise. If the other atoms were distributed clockwise, you would indicate it by writing the symbol twice – "@@". Issues to do with representing stereochemistry are discussed further in Chapter 8.2.

An extension of SMILES, called SMARTS, allows you to represent substructures. Most of the SMILES rules apply except that unsubstituted positions are no longer assumed to be filled by hydrogen atoms. Additional rules allow you to make generic statements about atoms and bonds in the substructure. So, for example, if "[O,N]" appears in a SMARTS string it means that the atom at that position in the structure is allowed to be oxygen or nitrogen, and "~" represents a bond that is allowed to be of any bond order.

A further extension, SMIRKS allows you to represent generic reactions. The reactant set and product set are each represented as SMILES/SMARTS strings (with some constraints on what features of the codes are allowed in SMIRKS). Each atom in the reactants which needs to be mapped to an atom in the products is numbered following a colon within the square brackets and the atom in the products is given the same number. The two strings are joined together with the symbol ">>". If you want also to specify a reagent, you include it between the two ">" symbols. SMIRKS is complicated and there are some problems of ambiguity in it because of clashes between what is needed in SMIRKS and what SMILES and SMARTS provide.

Other researchers, for example Richard Bone *et al*,[9] have described extensions to SMILES/SMARTS/SMIRKS to suit different applications.

5.3   CHMTRN and PATRAN

Some time before the development of SMARTS, the team working on LHASA had already developed a language called PATRAN[10] tailored to describing keying substructures (retrons and alerts) in synthesis planning and toxicity prediction systems, but a description of PATRAN has never been published in the open scientific literature. It was used in conjunction with a knowledge base programming language developed earlier in the LHASA project, CHMTRN, which I will describe first.

The spelling of the name of CHMTRN (pronounced "kemtran") echoes the practice of using six-character, upper-case names for variables in FORTRAN, typically omitting vowels, at the time when CHMTRN was developed. However, CHMTRN's style as a language was based on the same premise that produced the COBOL language at about that time – namely that a computer programming language should be similar in style to natural English language. Also, being of its time, CHMTRN is to a degree a procedural language. That is, it sets out a sequence of actions that the computer program is required to carry out. Being nevertheless also ahead of its time, it includes declarative elements. That is, it states conditions or facts and it is for the language interpreter and/or the program at run time to work out what the consequent actions should be and the order in which they should be carried out.

At a time when most scientific applications were hard-coded in FORTRAN by computer programmers, the LHASA knowledge base was kept separate from the LHASA program and it was written in a language, CHMTRN, intended to be used by a chemist not trained as a programmer. The intention was not fully realised, but it came close. The knowledge base was developed and maintained by chemists but, although they did not need to be trained programmers, they did need to understand and apply some programming concepts. The underlying procedural nature of the language means that while the outcome of writing things in a different order may be the same there can be big program performance differences that the skilled CHMTRN writer needs to be aware of. There are constraints on the construction of CHMTRN statements which the writer has to understand, connected with how they are stored by the CHMTRN compiler and interpreted by LHASA at runtime.

Figure 5.4 is a simplified example of what a transform keyed by two functional groups written in CHMTRN might look like. It is based on one written by Jan-Willem Boiten at the Catholic University in Nijmegen (now the Radboud University) but I have modified some parts and omitted quite a lot for the purposes of this illustration. Every transform starts with a number and name for the transform and information about publications on which the transform is based (the original lists more publications than I have kept in this illustration). Statements follow about the practical usefulness of the reaction to which the transform relates, from which the program generates a preliminary numerical rating for display to the user. This helps the program and the user to assess the relative merits of different routes in a synthetic tree. A picture of the retro-reaction follows for the convenience of people working on the knowledge base, and then the keying functional groups (the retron) are specified. Some qualifying statements follow, to prevent the transform from firing if there are unfavourable features in the query structure. Less calamitous features might have been dealt with by reducing the rating of the transform instead of killing it (*e.g.* "SUBTRACT 10 IF . . . " instead of "KILL IF . . . ") but there are no such rating-modification statements in this transform. So-called "mechanism commands" placed between lines containing four dots instruct the program on how to convert the target structure into the precursor and at the end typical conditions or reagents are given, which the program takes into account when considering the need to protect functional groups in other parts of the structure. If such protection is needed, the program automatically decrements the rating for the transform. Note the use of terms such as "WITHDRAWING BOND" and "LEAVING GROUP": the transform writer does not need to write or to call subroutines to recognise them. Less obvious, unless you are familiar with

```
      TRANSFORM 291
      NAME Reformatsky Reaction
...
...  REFERENCES
...  Synthesis, 571 (1989);
...  JACS 70, 677 (1948);
...  JOC 34, 3689 (1969).
...

      TYPICAL*YIELD                    FAIR
      RELIABILITY                      GOOD
      REPUTATION                       EXCELLENT
      HOMOSELECTIVITY                  FAIR
      HETEROSELECTIVITY                FAIR
      ORIENTATIONAL*SELECTIVITY        NOT*APPLICABLE
      CONDITION*FLEXIBILITY            FAIR
      THERMODYNAMICS                   EXCELLENT

...
...   O   OH       O       O
...   "   |        "       "
...   C-C-C    =>  C-C  +  C
...   |           |       |
...   OR,NR2      Br      OR,NR2
...

      ...Jan-Willem Boiten, Nijmegen, Apr. 1989
      ...Modified by P N Judson for use as an illustration only, Dec. 2008.
      ...PATH 2 BONDS
      GROUP*1 MUST BE ESTER OR AMIDE*3
      GROUP*2 MUST BE ALCOHOL

      ...
      KILL IF THERE IS A MULTIPLY BONDED ATOM AT ATOM*2 OR: AT ATOM*3
      KILL IF THERE IS A HETERO ATOM ALPHA TO ATOM*2
      KILL IF THERE IS ANOTHER WITHDRAWING BOND ON ATOM*2
      KILL IF ATOM*3 HAS A WITHDRAWING BOND
      KILL IF ALPHA TO ATOM*2 OFFPATH HAS A LEAVING GROUP

      ....
        BREAK BOND*2
        DELETE HETERO2*1
        ATTACH A CARBONYL TO ATOM*3
        ATTACH A BROMIDE TO ATOM*2

      ....
      CONDITIONS ORGANOZINC
```

Figure 5.4   An abridged example of what a two group transform in LHASA might look like.

CHMTRN, is the use of buzz words – words that are ignored by the program – to make the code easier to read and understand. "THERE", "IS", "A", "BONDED", "ATOM" and "AT" are all buzz words. "KILL IF THERE IS A MULTIPLY BONDED ATOM AT ATOM*2 OR: AT ATOM*3" could have been written as "KILL IF MULTIPLY ATOM*2 OR: ATOM*3", which would be very obscure compared with the better worded statement in the Figure 5.4.

A limitation with CHMTRN was that transforms could be keyed only by a set of sixty four, predefined functional groups. PATRAN removed this constraint, allowing the addition, for example, of heterocyclic chemistry to the knowledge base. Figure 5.5 is an extract from an example I have written for (retrosynthetic) conversion of a pyrazole to a diketone and a hydrazine to illustrate how CHMTRN and PATRAN work together. I have left the references and the rating information out of the figure to keep it short. It should not take you too much effort to work out what the code is about. For the sake of illustration, I have assumed that this synthetic method is unacceptable if the target structure is an N-substituted pyrazole and the substituents at ring positions 3 and 5 differ, on the grounds that a mixture will be generated in the forward direction. A real

```
        TRANSFORM 999
        NAME Pyrazole formation from diketone

    ...
    ...[references]
    ...[references]
    ...[rating statements]

    ...       ______          ______
    ...    ||        ||      ||        |
    ...    N      /    =>    O         \\
    ...      \   /                      O
    ...       N
    ...
    ...STARTP
    ...N%N%C[HETS=1]%C%C[HETS=1]%@1
    ...ENDP

    ...
    ...An asymmetrical diketone will lead to a mixture if there is a substituent on the N atom, so
    ...kill the transform: i.e. kill it if only one of atom*3 ...and atom*5 is substituted, or if both
    ...are substituted and the substituents are different
            IF THERE IS AN ATOM ALPHA OFFPATH FROM ATOM*1 &
              OR:IF THERE IS AN ATOM ALPHA OFFPATH FROM ATOM*2
            BEGIN BLOCK1
              IF THERE IS AN ATOM ALPHA OFFPATH FROM ATOM*3 &
                AND:IF THERE IS NOT AN ATOM ALPHA OFFPATH FROM ATOM*5 &
                THEN KILL
              IF THERE IS AN ATOM ALPHA OFFPATH FROM ATOM*5 &
                AND:IF THERE IS NOT AN ATOM ALPHA OFFPATH FROM ATOM*3 &
                THEN KILL
              SAVE AS 1 THE ATOM ALPHA OFFPATH FROM ATOM*3
              SAVE AS 2 THE ATOM ALPHA OFFPATH FROM ATOM*5
              KILL IF THE APPENDAGES FROM ATOM*3 TOWARDS &
                SAVED*ATOM 1 AND FROM ATOM*5 TOWARDS SAVED*ATOM 2 &
                ARE NOT IDENTICAL
            BLKEND BLOCK1
            ....
                BREAK BOND*2
                BREAK BOND*5
                ATTACH A KETONE ON ATOM*3
                ATTACH A KETONE ON ATOM*5
            ....
```

Figure 5.5   An abridged example of a pattern-based transform in LHASA.

transform would more likely allow asymmetrical substitution but apply a penalty to the rating, and it might deal with features in substituents at positions 3 and 5 that could interfere with the behaviour of the ketone groups in the precursor. It would also take into consideration the nature of the substituent at position 4 (consider the implications, for example, if it were an acyl group).

Atoms and bonds are automatically numbered from left to right in a PATRAN string. They can be referred to in CHMTRN statements as illustrated in Figure 5.5, allowing transforms to be written in great detail. A lot of information can be incorporated into the PATRAN string itself. Rings are defined differently from the way that they are defined in SMILES. As atoms are automatically numbered, there is no need to flag the first atom of two that are linked to close a ring. Instead, the bond that connects back is followed by "@" and the number of the atom to which it connects. The pattern in Figure 5.5 illustrates this, the "@1" at the end of the string indicating that the final bond links back to atom number 1, making a pyrazole ring. Also illustrated in Figure 5.5 is the way in which attributes can be attached to atoms. You will have worked out that "C[HETS=1]" means a carbon atom with (exactly) one heteroatom attached to it. It is similarly possible to specify the number of attached hydrogen atoms, atomic charge, whether the atom is in a ring and, if so, of what size, and so on. Bond attributes allow you to specify whether a bond is a fusion bond between two rings and, if so, whether the rings are aromatic, aliphatic, or can be either. An atom can be allowed to be one of several elements and a bond can be allowed to be of more than one order by using commas. For example "N,O-,=C" would match to amines, ethers, alcohols, imines, and ketones among other groups.

Later, so-called "2D patterns" were added as a further language development. The picture in Figure 5.4 is actually a valid 2D pattern. They were introduced during work on making it possible for LHASA to operate also in the synthetic direction, which requires transforms to describe the reacting centres in precursors explicitly as well as in products: if reactants are only generated by mechanism commands when transforms are applied, there is no keying information for the program to search for in the knowledge base when it is running in the synthetic direction. Reaction information could have been expressed by extending the "1D" pattern code but the LHASA project had a strong research element in it and the researcher was interested in exploring new ways of encoding information about structures and reactions. 2D patterns had the attraction of being more immediately understood by a chemist than 1D patterns or CHMTRN statements. 2D patterns could have rendered the description of the reaction by GROUP statements or by "1D" patterns redundant, but they were retained for compatibility with the existing knowledge base and to support program code that depended on them.

5.4 ALCHEM

The knowledge base language for the SECS program, ALCHEM,[11] looks rather like a combination of CHMTRN and PATRAN in a single language. It is close enough for someone familiar with CHMTRN and PATRAN to understand ALCHEM and translate it into CHMTRN and PATRAN, and *vice versa*. This is what we did when the groups working on PASCOP and LHASA exchanged some of the contents of their knowledge bases, as mentioned in Chapter 3.1.1, but I do not now have access to details about ALCHEM and a full specification for the language does not appear to have been published in the open literature.

5.5　Molfiles, SDfiles and RDfiles

For the purposes of exchanging chemical structural information between software applications, as distinct from communicating with a human user, a connection table in text file format developed at MDL Information Systems is widely used: the specifications for the Molfile format were published by Arthur Darby and coworkers[12] (the name was generally written "MOLfile" but is now more usually written as "Molfile"). Specifications for updated versions are published from time to time by MDL Information Systems (now owned by Symyx). Figure 5.6 shows a Molfile for furfural, Structure 5.2 on page 51, in which the atom numbers appearing in the Molfile are shown. Software from many suppliers will write Molfiles but I created this one with ISIS Draw since both originate from the same source – MDL Information Systems, now Symyx. The Molfile starts with a header and a comment line (empty in the example in Figure 5.6). A blank line follows, and then a line containing various pieces of general information. In this example the only information relevant to us is the number of atoms and number of bonds at the start of the line. The next section is a table giving the coordinates of atoms in the structure and their chemical

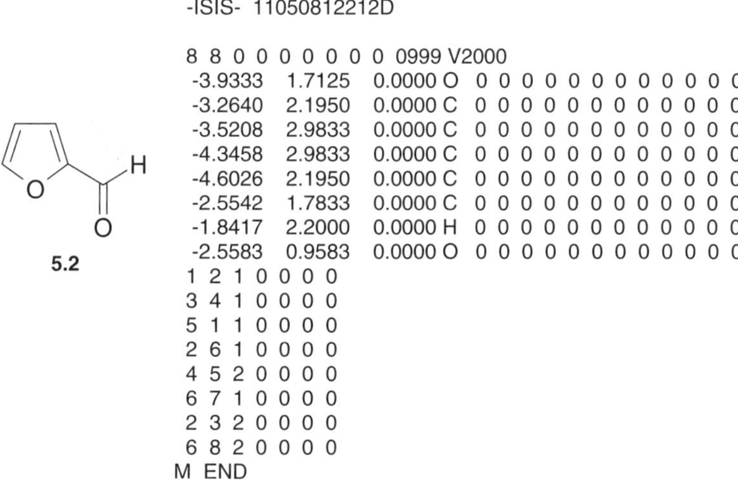

Figure 5.6　Furfural and a molfile for it.

symbols. This Molfile describes a standard chemical diagram and so all of the z coordinates are zero (though in this case they might have been zero anyway depending on the angle from which the molecule was viewed, since it has a planar structure). After the atom table there is a bond table. So, for example, the first line shows that atom 2 is joined to atom 3 by a double bond. After that comes the END statement.

5.2

An SDFile is a set of Molfile format connection tables all written into one file. The tables are separated by lines containing "$$$". The specification allows associated data to be recorded after each connection table, each line starting with a code to identify the data item which might be, for example, a physical property or activity in a biological assay. An extension of the format allows information about reactions to be stored in RDfiles. Storing information about reactions is complicated by the need to record which atoms in the starting materials correspond to which in the products – a topic which is mentioned also in Chapter 3.2.

5.6   Mol2 files

There are several widely used file formats for exchanging data between molecular modelling and protein structure determination software but they are outside the scope of this book. The Mol2[13] format from Tripos is worth mentioning because it is favoured by some people working with chemical information outside the field of molecular modelling and because it is rather similar to Molfile format but with more flexible formatting and broader coverage of data beyond the description of a chemical structure. The approach to dealing with the graphical information is similar, with a table for atoms and a table for bonds, although there are differences of detail. The overall format of the file is less rigid because each section in it is identified by a header, allowing some flexibility in the layout of the file. Mol2 having been developed in connection with molecular modelling, there are specific headers for many molecular properties.

5.7   The Standard Molecular Data Format and Molecular Information File

An international effort over several years in the 1990s aimed to provide a standard format for the exchange of chemical structural and related data.[14,15] Originally termed the Standard Molecular Data Format (SMD) and later the Molecular Information File (MIF), it was intended to be non-proprietary and it introduced the idea of communicating data in blocks (as it is in Mol2 and, for example, more flexibly in XML files) rather than in fixed format (as it is in Molfiles and SDfiles). By way of examples, a NODE block contained a list of the atoms in a structure; a CONVENTION block, divided into sub-blocks, specified bonds, hydrogen counts for each atom, charges, and radical states; a COORD block contained 3D coordinates for the structure; a DISPLAY-COORD block contained 2-D coordinates for graphical display. The specification allowed blocks to be in any order and allowed any block to be included or omitted.

SMD/MIF seems to have foundered, like FORTRAN 99, because the slow committee processes that go with this kind of project allowed it to be overtaken by urgent necessity. Molfile, SDfile, and RDfile formats were well on the way to becoming *de facto* standards even before MDL Information Systems published them and renounced copyright constraints on their use,[12] and SMD/MIF was never widely taken up in their place.

5.8   Chemical Markup Language

Chemical Markup Language, CML, as its name implies, is allied to XML but designed for the communication of chemical information. First described nearly a decade ago,[16–18] it has undergone steady development championed especially by Peter Murray-Rust at the University of Cambridge and Henry Rzepa at Imperial College, London. It goes far beyond providing for communication of the structures of small molecules. Recent publications, for example, cover the representation of spectral data[19] and polymeric structures.[20] You can find a great deal about CML *via* the worldwide web – not surprisingly, given that the first intended purpose of CML is to facilitate better communication about chemistry over the web. A Google search for "CML" together with "Rzepa" or "Murray-Rust" is a good way into it.

5.9   Using Pictures

All sorts of other non-graphical representations of chemical structure have been devised for particular pieces of software. Some of them are neat and convenient, many are anything but. None of them have been widely adopted. SMILES, and its derivatives, and Molfiles, and their derivatives, are currently the *de facto* standards for chemical information systems although CML and InChI (see Chapters 5.8 and 6.1.5) may grow in popularity – the first for communication of more complex data than just basic chemical structure information and the second because of the flexibility it offers in the level of detail about chemical structures that you choose to record and/or to search for.

Many applications that depend on SMILES do so for historical reasons. They were designed when users did not have access to simple graphics facilities and the legacy lives on. Such systems usually offer an optional graphics front end – sometimes their own and sometimes a third party one such as Marvin.[21] The user draws the structure of interest and the add-on package generates a SMILES string which is passed to the program in the background. Most independent chemical structure editors can also generate molfiles for communication with applications that require them as input. The major chemical database and knowledge-based systems use structural diagrams as their primary means of communication with human users, either *via* modules built specifically for the systems or *via* automated links to third party structure editors. Knowledge-based systems such as Derek for Windows, which will be described in Chapter 13, have graphical interfaces for knowledge base development as well as for communication with end users.

There will continue to be uses for non-graphical codes to represent chemical structures but, in words worn thin in the re-telling, a picture is worth a thousand words. Most of the current uses of non-graphical codes in communications with human users are consequences of history and we can hope that chemists, at least, can look forward to being able to interact with computer systems by using graphics as a matter of course. Time was when there was no alternative to sitting down with a piece of paper and working out the IUPAC chemical name for a structure in order to be able to look it up in a printed index. There seems little justification for obliging chemists to sit down and work out linear codes instead. That time is past – computers can deal directly with chemical structures.

References

1. W. J. Wiswesser, in *A Line-Formula Chemical Notation*, Crowell, New York, USA, 1954.
2. E. G. Smith, in *The Wiswesser Line-Formula Chemical Notation*, McGraw-Hill, New York, 1968.
3. G. Palmer, Wiswesser Line-Formula Notation, *Chem. Britain*, 1970, **6**, 422–426.
4. J. J. Vollmer, Wiswesser Line Notation: an Introduction, *J. Chem. Educ.*, 1983, **60**, 192–196.
5. D. Weininger, SMILES, a Chemical Language and Information System, *J. Chem, Inf. Comput. Sci.*, 1988, **28**, 31–37.
6. D. Weininger, SMILES 3. Depict. Graphical Depiction of Chemical Structures, *J. Chem, Inf. Comput. Sci.*, 1990, **30**, 237–243.
7. D. Weininger, SMILES 2. Algorithm for Generation of Unique SMILES Notation, *J. Chem, Inf. Comput. Sci.*, 1989, **29**, 97–101.
8. http://www.daylight.com/.
9. R. G. A. Bone, M. A. Firth and R. A. Sykes, SMILES Extensions for Pattern Matching and Molecular Transformations: Applications in Chemoinformatics, *J. Chem. Inf. Comput. Sci.*, 1999, **39**, 846–860.
10. G. A. Hopkinson, *Computer-Assisted Organic Synthesis Design*, PhD thesis, University of Leeds, 1985.

11. W. T. Wipke, Computer-assisted Three Dimensional Synthetic Analysis, in *Computer Representation and Manipulation of Chemical Information*, ed. W. T Wipke, S. R. Heller, R. J. Feldmann and E. Hyde, Wiley, New York, 1974, pp. 147–174. (ALCHEM is described on pp. 165–168).
12. A. Dalby, J. G. Nourse, W. D. Hounshell, A. K. I. Gushurst, D. I. Grier, B. A. Leland and J. Laufer, Description of Several Chemical Structure File Formats Used by Computer Programs Developed at Molecular Design Limited, *J. Chem. Inf. Comp. Sci.*, 1992, **32**, 244–255.
13. The Mol2 format was developed by Tripos L.P., St Louis, Missouri, USA, and the specification for it can be downloaded from http://www.tripos.com.
14. F. H. Allen, J. M. Barnard, A. P. F. Cook and S. R. Hall, The Molecular Information File (MIF): Core Specifications of a New Standard Format for Chemical Data, *J. Chem. Inf. Comput. Sci.*, 1995, **35**, 412–427.
15. J. M. Barnard, The Standard Molecular Data (SMD) Format, in *Chemical Structures 2: The International Language of Chemistry*, ed. W. A. Warr, Springer-Verlag, Berlin, Heidelberg, 1993, pp. 185–193.
16. P. Murray-Rust and H. S. Rzepa, Chemical Markup Language and XML Part I. Basic Principles, *J. Chem. Inf. Comp. Sci.*, 1999, **39**, 928–942.
17. P. Murray-Rust, H. S. Rzepa, M. Wright and S. Zara, A Universal Approach to Web-Based Chemistry Using XML and CML, *Chem. Commun.*, 2000, 1471–1472.
18. P. Murray-Rust, H. S. Rzepa and M. Wright, Development of Chemical Markup Language (CML) as a System for Handling Complex Chemical Content, *New J. Chem.*, 2001, **25**, 618–634.
19. S. Kuhn, P. Murray-Rust, R. J. Lancashire, H. S. Rzepa, T. Helmusk, E. L. Willighagen and C. Steinbeck, Chemical Markup, XML, and the World Wide Web. 7. CMLSpect, an XML Vocabulary for Spectral Data, *J. Chem. Inf. Model.*, 2007, **47**, 2015–2034.
20. N. Adams, J. Winter, P. Murray-Rust and H. S. Rzepa, Chemical Markup, XML and the World-Wide Web. 8. Polymer Markup Language, *J. Chem. Inf. Model.*, 2008, **48**, 2118–2128.
21. Marvin is supplied by ChemAxon Kft., Máramaros köz 3/a, Budapest, 1037 Hungary. http:// www.chemaxon.com.

CHAPTER 6

Structure, Sub-Structure and Super-Structure Searching

This is not, for the most part, a book about the inner workings of chemical information packages, but some background on topics that impact upon the functioning of knowledge-based and similar systems may be useful. This chapter does not cover, by a long way, all the work that has been published on methods for searching in chemical structure and reaction data bases. I have just tried to illustrate the main approaches.

6.1 Exact Structure Searching

Searching for complete matches between a query structure and candidate structures in a database is usually termed exact structure searching. The usage of the word, "exact", is not a strict one. In most chemical structure database systems the user has some flexibility over whether an "exact" search should return only stereochemically identical matches or should return all stereo-isomers of the query. Similarly, there is normally flexibility about whether to take isotopic differences into consideration.

To be able to search reliably for something – in this case, a specific chemical structure – you need a way of identifying it that is unique and unambiguous. A name or code is unique if it is the only one that can be generated for a given structure. It is unambiguous if it defines only one structure. CH_4 as a representation for methane meets both criteria, as long as you apply the normally accepted conventions for ordering the elements in a molecular formula (in the absence of those conventions, CH_4 would not be a unique name for methane, because its formula might be represented as H_4C). However, in general, molecular formulae are not unambiguous. C_4H_{10}, for example, might be *n*-butane, or *iso*-butane (2-methylpropane).

RSC Theoretical and Computational Chemistry Series No. 1
Knowledge-based Expert Systems in Chemistry: Not Counting on Computers
By Philip Judson
© Philip Judson 2009
Published by the Royal Society of Chemistry, www.rsc.org

IUPAC nomenclature rules are not rigorous enough, or are not applied rigorously enough, to provide unique and unambiguous names. Even if they were, the names would be a poor choice for fast searching by computer because they would be cumbersome to store and to search. Chemical Abstracts Service numbers (CAS Numbers) are ideal for fast searching, because they require only integer searching. They are also very compact, which makes it possible to hold large lists of them in computer memory. They do not fully meet the ideals of being unique and unambiguous. There are structures that have more than one CAS number (sometimes because different macroscopic forms of a compound have different numbers and sometimes because numbers were originally assigned to what were believed to be two substances but later recognised to be one) and there are examples of CAS Numbers that represent more than one structure (*e.g.* 1330-20-7 is the CAS number for mixed isomers of xylene). However, they are sufficient for many purposes and they are widely used.

But there is a fundamental problem with CAS numbers when it comes to computer processing of structures: there are no algorithms than can generate a CAS number from a structure or *vice versa* and there cannot be, because there is no formal correspondence between them. CAS numbers are simply assigned. If you want to be able to enter search queries as structural diagrams you need an algorithm to generate unique and unambiguous representations of structures suitable for storing and fast searching in a computer. Connection table representations such as the ones recorded in Molfiles (see Chapter 5.5) should be unambiguous, as long as they are complete, and rules could be devised to make them unique, but they are not attractive candidates for fast computer searching. What you need is a number or a simple character string.

Wiswesser Line-formula Notation (WLN – see Chapter 5.1) is one possible solution, and indeed that was one of the reasons for its development, but it is found now only in a few early databases that are still in use. Canonical SMILES codes provide another solution. Methods in widespread use are variants of the Morgan algorithm and SEMA, a stereochemical extension of it but several other methods have been described and are used.

6.1.1 Canonical SMILES Codes

David Weininger set out rules for generating a unique SMILES code for a structure.[1] The first phase of the process ("CANON") is to rank the atoms in the structure as follows.

Step 1: List the following attributes for each non-hydrogen atom in the structure:
1. number of connections
2. number of non-hydrogen bonds
3. atomic number
4. sign of charge
5. absolute charge
6. number of attached hydrogen atoms.

Step 2: Rank and number the atoms according to the above attributes, sorting first on attribute 1 and then, if necessary, on attribute 2 and so on.

Step 3: Replace the ranking numbers with prime numbers, starting from 2 (atoms ranked first become numbered 2, atoms ranked second become numbered 3; atoms ranked third become numbered 5; atoms ranked fourth become number 7; and so on).

Step 4: Replace the number for each atom with the product of the numbers currently associated with its neighbours. [Replacing ranking numbers with prime numbers in Step 3 and then multiplying them together in Step 4, rather than just adding together the ranking numbers was proposed by Malcolm Bersohn [2] as a way of avoiding ambiguities that arise with simple summation (*e.g.* if two atoms had three neighbours ranked 1, 4, 4 and 2, 2, 5 the sum for each of them would be 9)].

Step 5: Re-rank the atoms consecutively starting from one, in the order set by the results of Step 4.

Step 6: Cycle through Steps 3 to 5 until either the highest ranking is equal to the number of atoms (*i.e.* every atom is uniquely ranked) or there is no further change in the ranking.

Step 7: If the highest ranking is not equal to the number of atoms, there is symmetry and so in order to settle upon a single complete ranking set that can be used for further computer processing, double the ranking numbers for all the atoms, choose the first of the highest ranking pair of atoms with equal ranking and reduce its ranking by 1. Reorder the ranking of all the atoms consecutively from one and go back to Step 6. Continue like this if necessary, until every atom is uniquely ranked.

Table 6.1 illustrates the process for 1-methyl-5-ethylpyrazole, Structure 6.1, which has no problem symmetry and provides a complete ranking without the need to apply Step 7. It is worth commenting on the first two attributes listed for Step 1. The "number of connections", the first attribute, is the number of connections in the graph, which normally means the number of adjacent non-hydrogen atoms, since the hydrogen atoms are not included in the graph. The number of connections to atom 1 in Structure 6.1 is 1, not 4, for example. The second attribute, "number of non-hydrogen bonds" does not mean the

Table 6.1	Applying the ranking procedure to find the canonical SMILES string for Structure 6.1.

Atom No.	*1*	*2*	*3*	*4*	*5*	*6*	*7*	*8*
Step 1	1, 1, 6, 0, 0, 3	2, 2, 6, 0, 0, 2	3, 4, 6, 0, 0, 0	2, 3, 6, 0, 0, 1	2, 3, 6, 0, 0, 1	2, 3, 7, 0, 0, 0	3, 3, 7, 0, 0, 0	1, 1, 6, 0, 0, 3
Step 2	1	2	6	3	3	4	5	1
Step 3	2	3	13	5	5	7	11	2
Step 4	3	26	165	65	35	55	182	11
Step 5	1	3	7	6	4	5	8	2

"number of bonds that are not hydrogen bonds" and it does not mean the "number of connections to adjacent atoms that are not hydrogen": the word "bonds" is used here in the traditional valency sense where, for example, a double bond counts as two bonds. So the value of the second attribute for atom 3 in Structure 6.1 is 4, and for atom 8 it is 1, for example.

6.1

The second phase of the process ("GENES") is to generate a SMILES string for the structure, using the atom rankings to decide where to start and in which direction to go first at branches. The path is grown depth first – *i.e.* when you come to a branching point you proceed along one branch as far as possible, doing the same at any succeeding branching points that you encounter, before stepping back to apply the same procedure to the next ranked branches until the whole structure has been traversed. Two extra rules are needed to take account of rings:

1. Branch in a ring towards a double or triple bond if there is one attached to the branching point; if not follow the normal rule of branching towards the highest ranking atom (*i.e.* the one labelled with the smallest number).
2. If you encounter a node that is already in the code that has been generated so far, stop growing along that path (you have found a ring closure point).

So the SMILES code for Structure 6.1 starts with C for atom 1, then C for atom 2. At this point we have a decision to make which exposes a weakness in SMILES codes. Shall we treat the pyrazole ring as an aromatic one or as non-aromatic. Two things hang upon the decision: if the ring is aromatic, then the symbol for atom 3 is "c" and the next atom in the string will be atom 4, that being higher ranked than atom 7; if the ring is non-aromatic, then the symbol for atom 3 is "C" and the next atom in the string will be atom 7, since, in accordance with the extra rule 1, we should branch towards the double bond. Continuing the same selection process, the complete canonical SMILES string is CCc1ccnn1C if we treat the ring as aromatic, and CCC1=N(C)NC=C1 if we treat it as non-aromatic. In a recent paper, Neglur, Grossman, and Liu[3] also point out that the CANON and GENES algorithms may not generate a unique code because of ambiguities associated with representations of aromaticity. For example, you can get either of two SMILES codes for 3,5-diethyl-methylbenzene (3,5-diethyltoluene), Structure 6.2 – CCC1= CC(=CC(=C1)C)CC or CCC1=CC(=CC(=C1)CC)C – if you use an alternating bond representation for the ring, rather than aromatic bonds.

6.2

In practice this is not normally a problem. The code is not intended to support direct searching across different computer applications. Canonical codes are generated internally by applications and stored to support fast searching. So as long as the application is consistent within itself that is all that matters. The problem with encoding aromatic rings does not arise in a well designed application because rules will be built into it about when to define a bond as aromatic, when as double or single; if aromatic rings are represented with localised double and single bonds in communications with the user there can still be rules about treating them as aromatic for the purposes of generating canonical SMILES codes.

There is another reason why canonical SMILES codes are not suitable as unique codes to support searching across different applications. There are improvements to the rules for creating SMILES from time to time. Codes created by one version of the algorithm are not always the same as those created by another. So even if two applications start off using compatible SMILES codes, this may not remain the case.

6.1.2 Morgan Names and SEMA Names

Working for Chemical Abstracts Service, Harry Morgan[4] developed an algorithm intended to generate unique and unambiguous identifiers for chemical structures. The resultant identifiers cannot generally be reverse engineered to recreate the original structures but that does not matter – they provide what is needed for reliable, fast, exact searching in a computer database. The Morgan algorithm was subsequently extended to include stereochemical information by Todd Wipke and Thomas Dyott[5] to generate what is generally known as the "SEMA name" for a structure, the term which they proposed derived from "Stereochemical Extension of the Morgan Algorithm". Current database systems use an assortment of variations on these methods but they still tend to be referred to informally sometimes as the Morgan Algorithm, sometimes as SEMA.

In his paper, Morgan first describes how to generate a set of codes, in the form of character strings, that are unambiguous – *i.e.* can only have been generated from the one structure. Depending on where you start in the structure and which way you progress around it there will be many possible codes. In theory, you could generate all of them and then apply ranking rules to decide which one to select as the unique identifier, but that could involve processing hundreds or even thousands of strings. In the second part of his paper, Morgan describes a procedure for the preliminary ordering of the atoms before generation of an unambiguous code to circumvent this problem. The following description instead starts with the ordering of the atoms. It is the way the algorithm was introduced to me and it is also broadly the way it is described in the Encyclopedia of Computational Chemistry.[6]

The first step is to label the atoms in the structure according to their degrees of connectivity (to non-hydrogen atoms), as illustrated in Structure 6.3a.

Re-label the atoms with the sums of the values in their neighbours' labels, as illustrated in Structure 6.3b. Count how many different labels you have. Iterate the process of re-labelling by taking the sums of the values of neighbours until either every atom has a unique label or the number of different labels no longer increases. If the second case applies, retain the labels you had before the final iteration (that being the one either that has a bigger number of different labels than the last one or that has the same number but based on smaller values). Morgan calls these labels the EXTENDED CONNECTIVITY values. In the case of the pyrazole we are working on, the number of different labels in Structure 6.3c is the same as in Structure 6.3b and so the process stops and the labelling of Structure 6.3b is retained. Now re-label the atoms sequentially from 1 as follows. Start with the atom that has the biggest label in Structure 6.3b followed by its neighbours in order from the highest to the lowest labelled. Repeat the process for atoms so-far not renumbered that are attached to the neighbour to atom number 1 that was highest labelled, and so on, to produce the ranking in Structure 6.3d (if there were symmetry in a graph, you would come upon cases where you had to choose between atoms with the same EXTENDED CONNECTIVITY values, in which case you would create a set of sequentially labelled structures – one for each option).

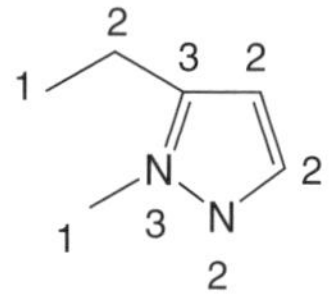

6.3a

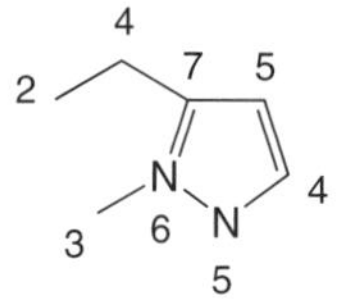

6.3b

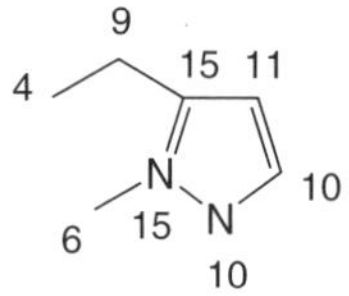

6.3c

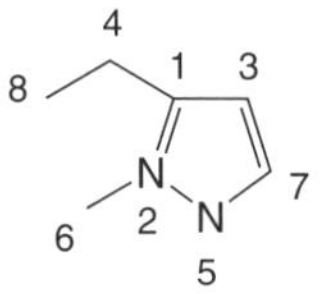

6.3d

The next step is to create four lists.

Morgan calls the first list the "FROM ATTACHMENT" list, but it is often referred to more succinctly as the "FROM" list. In the implementation described by Morgan, atom numbers are given leading zeros in this list so that they are all three figure numbers (*e.g.* "1" becomes "001"), but for the sake of simplicity the leading zeros are omitted in the following description. For each atom in order, list the number of the lowest numbered atom attached to it unless that atom is higher numbered than the current one, in which case make no entry in the list. For Structure 6.3d this gives

FROM: 1 1 1 2 2 3 4

There is a no entry for atom 1, since it is not connected to any atom with a lower rank. Atoms 2, 3, and 4 are all connected to atom 1, atoms 5 and 6 are connected to atom 2, atom 7 is connected to atom 3 and atom 8 is connected to atom 4.

The way in which the FROM list is created means that one bond in every ring in a structure will be absent. So a second list is written – the RING CLOSURE list. Each line in the list contains the numbers of the two atoms for which a connection is not included in the FROM list. There is one ring in Structure 6.3d and hence one missing connection:

RING CLOSURE: 5 7

Next comes a list of atom types, defined by their elemental symbols, termed "NODE VALUES" by Morgan. It is slightly surprising that, for a computer identification code, the textual symbols were preferred to numerical atomic numbers but Morgan does not comment on the reason. Finally, comes the "LINE VALUE" list – a list of the types of the bonds defined by the FROM and RING CLOSURE lists.

NODE VALUES: C N C C N C C C

LINE VALUES: 1 2 1 1 1 1 1 2

All that remains is to concatenate the lists into a single string, reintroducing leading zeros for the numbers in the FROM and RING CLOSURE lists:

001001001002002003004005006CNCCNCCC12111112

In a case where there was symmetry in the graph there would be more than one such list. The rule for choosing the unique one is simple. It is the one that comes first in an alphanumeric listing of the possible completed codes. The practical implementation is a bit more cunning. Given the way that the code is built, you can compare the ranking of alternative codes stage by stage. As soon as one code ranks below another you can drop it. So for maximum processing efficiency the best way to go about the job is to start on two of the candidates coming from the original ranking of the atoms. As soon as one code can be ruled out, drop it and start building another one to compare with the winner, and so on. That way, you only have to complete the process for relatively few candidates.

Just as in the case of the algorithm for generating unique SMILES strings, the Morgan algorithm will give different answers for the pair of diagrams that can be drawn for a structure with alternating double and single bonds in an aromatic ring. You can still generate a unique code by generating the strings for both possibilities and applying Morgan's rule that makes the code coming first in an alphanumeric listing the unique one.

The SEMA name introduced by Todd Wipke and Thomas Dyott[5] is constructed by adding a "DOUBE BOND CONFIGURATION" list and an "ATOM CONFIGURATION" list to the Morgan string. These lists contain values for every double bond and for every atom in the structure, respectively, as follows: 0 for a non-stereocentre; 1 for an "odd" stereocentre; 2 for an "even" stereocentre; and 3 for a stereocentre of unknown configuration. Whether the parity of a tetrahedral stereocentre is odd or even for a given atom in the FROM list is determined as follows (determining the parity for the stereochemistry at a double bond follows an analogous procedure). Imagine you are looking down the bond towards the current atom of interest from the atom that has smallest Morgan EXTENDED CONNECTIVITY value out of the four attached to it. Taking the EXTENDED CONNECTIVITY values for the other three atoms in order, do they run clockwise or anticlockwise? If they run clockwise, the parity for the atom is classed as "even". If they run anticlockwise, the parity is classed as "odd".

Why "even" and "odd"? The terms come from the algorithm for working out whether the ordering of the EXTENDED CONNECTIVITY values runs clockwise or anticlockwise by computer. Consider a set of four atoms, 'b', 'c', 'd', and 'e', connected to a tetrahedrally-asymmetric atom 'a', in which atom 'b' has the smallest EXTENDED CONNECTIVITY value out of 'b', 'c', 'd', and 'e'. Imagine you are looking down the bond from atom 'b' towards atom 'a'. Write down the letter 'b' followed by the letters for the other atoms attached to 'a', selecting them in clockwise order. For one chirality your solution will be one of 'bcde', 'bdec', or 'becd'; for the other, your solution will be one of 'bdce', 'bced', or 'bedc'. Looking down the same bond and listing the atoms on the basis of their EXTENDED CONNECTIVITY values in accordance with the previous paragraph will have given you one of those six possibilities. To find out whether the list you got from the EXTENDED CONNECTIVITY values runs in the same order as the one found by viewing the 3D structure, set about converting the one into the other by swapping pairs of letters in the list. If the number of swaps required is zero or even, then the parity is even. If the number of swaps required is odd, then the parity is odd. For example:

to get from bcde to bcde:	bcde	= bcde	0 swaps
to get from bcde to bdce:	bcde –> bdce	= bdce	1 swap
to get from bcde to bdec:	Bcde–> bdce –> bdec	= bdec	2 swaps

and so on.

If you try all the possibilities you will find that whenever ordering by EXTENDED CONNECTIVITY values goes clockwise the parity comes out as even and whenever ordering by EXTENDED CONNECTIVITY values goes anticlockwise the parity comes out as odd.

Two further modifications to the process of generating a Morgan string are needed. Firstly, you need to distinguish between atoms that are and are not centres of asymmetry expediently. The Morgan algorithm provides a convenient way to do this because if two atoms are found to have the same EXTENDED CONNECTIVITY value the algorithm grows the corresponding branches fully in order to decide which one should have priority. If that process fails to find any difference, there must be symmetry or, in other words, the branch point cannot be a centre of asymmetry. Secondly, to ensure that the SEMA name will be unique you must incorporate the task of generating the double bond and atom configuration lists into the incremental process for building the Morgan string – you cannot generate a final Morgan string and then set about determining the stereochemical information and appending it.

6.1.3 MOLGEN-CID

The authors of the MOLGEN chemical identifier (MOLGEN-CID)[7] reported in 2004 that it could be accessed free of charge *via* the worldwide web, but I have not been able to find it in 2009. Perhaps I have not hit on the right search criteria or perhaps it is no longer available *via* the web. As described here it is intended for connected graphs and hence for covalently bonded structures. The first steps in the process of developing a canonical name for a structure are as follows:

1. rank all the non-hydrogen atoms in reverse order of their atomic numbers (*i.e.* with the biggest number first);
2. sub-rank each group of atoms with the same atomic number (a) by atomic mass if some of the atoms are different isotopes, (b) by charge if non-zero, (c) giving higher rank to atoms carrying an unpaired electron, and (d) having a valency differing from the default defined in the scheme (*e.g.* four for carbon);
3. sub-rank atoms that are still not distinguished according to whether they are ring atoms or chain atoms;
4. if there are chain atoms that are still not distinguished, sub-rank them according to whether they are in chains that link rings together or not;
5. rank all remaining undistinguished atoms according to the number of aromatic, triple, double, or single bonds connected to them, not counting bonds to hydrogen atoms.

Note that Step 5 means the MOLGEN algorithm shares a weakness with the one for generating canonical SMILES codes – namely, that it requires categorical decisions to be made about whether bonds are aromatic or not.

At the end of Step 5 there will probably still be some atoms that have not been uniquely ranked. You look at each of those that has been ranked uniquely, in order, and see if it is bonded to any of the atoms in the unresolved groups. If it is, then the atom bonded to it is given the highest ranking within its group. According to the authors, you often achieve a unique ranking for the whole set of atoms by the end of this process. If not, you enter a complicated

Table 6.2 Applying the ranking procedure to find the MOLGEN chemical identifier for Structure 6.1.

Step 1	6 7		1 2 3 4 5 8					
(Step 2)	6 7		1 2 3 4 5 8					
Step 3	6 7		3 4 5			1 2 8		
(Step 4)	6 7		3 4 5			1 2 8		
Step 5	7	6	3	4 5		2	1 8	
Refine based on 7	7	6	3	4 5		2	8	1
Refine based on 6	7	6	3	5	4	2	8	1

backtracking process in which you try giving higher ranking to each atom in an unranked set in turn and apply rules to decide which solution to keep, but I leave you to get the bottom of how this works by studying the publication, where you will find illustrative examples.[7]

Table 6.2 shows how the rules work for Structure 6.1 (assuming that the ring is an aromatic one). In this case, a unique ordering is reached after two stages of the refinement process that comes after Step 5.

6.1.4 The Method Described by Henrickson and Toczko

James Hendrickson and Glenn Toczko described a method for generating a unique numbering for the skeleton of a chemical structure.[8] In principle, it can be adapted to encode a structure in complete detail, with its atom and bond types, although they do not illustrate this fully in their paper. Conceptually the method is simple and elegant. A graph can be fully described by an adjacency matrix – *i.e.* a matrix a bit like a mileage chart in a road atlas – in which each element has the value 1 if the two nodes are connected and 0 if not (this representation of graphs came up also in Chapter 3.2). Only half of the matrix needs to be recorded, since it is symmetrical about the diagonal, and the diagonal itself can be omitted, since the elements in it would represent the connection of each atom to itself. The skeleton of structure 6.1, for example,

1	2	3	4	5	6	7	8	
-	1	0	0	0	0	0	0	**1**
-	-	1	0	0	0	0	0	**2**
-	-	-	1	0	0	1	0	**3**
-	-	-	-	1	0	0	0	**4**
-	-	-	-	-	1	0	0	**5**
-	-	-	-	-	-	1	0	**6**
-	-	-	-	-	-	-	1	**7**
-	-	-	-	-	-	-	-	**8**

Figure 6.1   An adjacency matrix defining the skeleton of Structure 6.1.

can be represented by the matrix in Figure 6.1. To generate a binary number to represent the structure you list the rows of the matrix consecutively from top to bottom.

These are not the only matrix and number you could get, since the numbering of the atoms was an arbitrary choice. However, one of all the possible numbers to represent the graph must be the biggest, and Hendrickson and Toczko define that as the unique identity number. If there are n atoms in a structure there will be $n!$ ways of numbering them and hence $n!$ alternative matrices (40,320 for Structure 6.1) and so it would be computationally inefficient to generate all of them and then to select the biggest. Obviously, the number generated from the matrix with the most ones at the start of the first row must be bigger than any other, since they represent the highest powers of two in the binary number. So the trick is to choose the atom with the greatest connectivity in the structure as atom 1. If there are several such atoms, you start with several tentative first rows. You repeat this process to make your choice of atom 2 for row 2, and so on. As you go along, some of your earlier alternatives will be ruled out as others overtake them in magnitude, and so getting to the end of the computation requires you to retain relatively few alternatives as the matrix grows and you eventually end up with one complete one, instead of a set of thousands to choose between.

This method for generating a unique number is closely similar to one published in a series of papers by Milan Randić *et al.*[9] but they defined the unique number to be the smallest one generated, rather than the largest. Hendrickson and Toczko realised that defining it as the largest makes it easier to eliminate the "wrong" matrices early as you build them row by row.

Hendrickson and Toczo suggest that information about the attributes of the atoms in a structure, such as what elements they are and/or the number of π bonds attached to them, could be stored in the diagonal of the matrix (the elements of the matrix representing the connection of each node to itself, so to speak) and appended to the binary identification number for the graph. Provided that the atoms are fully described, there will again only be one, biggest complete binary string to represent the structure.

6.1.5 InChI Code

The International Union of Pure and Applied Chemistry (IUPAC) is developing a code to represent chemical structures that is compatible with computer text searching systems, InChI (IUPAC International Chemical Identifier).[10] It is intended to be globally consistent whereas, as already mentioned, unique SMILES strings, Morgan strings, and SEMA names only guarantee internal consistency in an application. Having a fast method for searching globally has become of greater interest with the advent of the worldwide web and services such as Google. A user may want to search for an exact stereoisomer and/or tautomer or for all isomers and tautomers of a compound, and the isotopic composition of the compound might or might not matter. So an InChI code is divided into "layers" describing the structure in increasing detail. Layers other than the fundamental one defining chemical composition can be included or omitted to support encoding and searching at different levels of exactness. At the most general, it is possible to define and search for a chemical on the basis just of its molecular formula. At an intermediate level complete sets of tautomers, for example, have the same code. At the most exact level, a specific tautomer is defined, along with isotopic composition and stereochemistry. InChI codes are designed to be unique. Within their intended scope they might be described as unambiguous although that is not a very satisfactory use of the word: if, for example, the code is at the general tautomer level it represents a set of tautomers, not one, but it is unambiguous in the sense that it is the one set having the formula and connectivity defined by the code.

InChI codes are intended to be generated and interpreted by a software algorithm rather than by human users. Open Source software to generate InChI code from structures is available,[10] chemical drawing packages such as ChemDraw[11] and Marvin[12] can generate InChI codes, and there are websites where you can generate InChI codes online (*e.g.* the ACD Labs[13] and PubChem[14] websites) and sites where you can check the validity of InChI codes, do searches using InChI and make conversions both ways between InChI and Molfile formats (*e.g.* the ChemSpider[15] website). A current weakness of the InChI scheme, in my view, is that the only recognised valid InChI code for a structure is the one generated by the specific executable program authorised by IUPAC. In itself, that might make sense, but the formal specification and definition of the algorithm for generating InChI codes are not included with the Open Source documentation and have not been published. Perhaps there is a plan to address this shortfall but in the meantime, the InChI code generator does not meet a basic principle of good scientific practice, namely that other parties should be able to reproduce work independently.

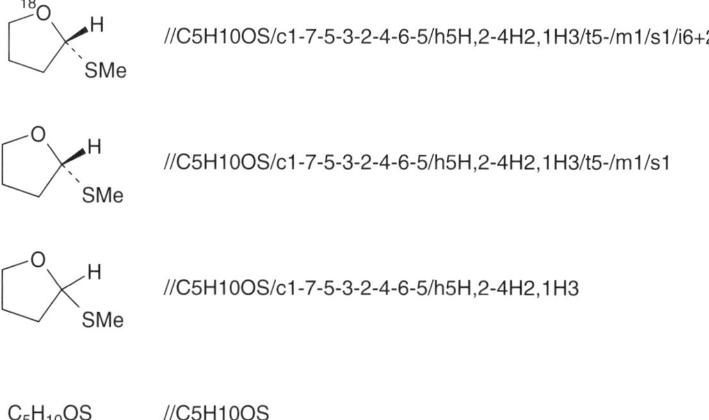

Figure 6.2   How InChI codes define structures in different levels of detail.

The examples in Figure 6.2, based on illustrations on Nick Day's website,[16] give an idea of what InChI codes look like. It is obvious how InChI deals with the molecular formula. Working out how it specifies atom types at the nodes in the chemical graph may take a bit of thought. If you want to get to the bottom of how it deals with stereochemistry you will need to do some research, but it is easy to see how it deals with isotopes.

6.2 Atom by Atom Matching

The purpose of generating unique codes by the methods described in the first part of this chapter is to speed up database searching. You can only guarantee an exact match if the codes are also unambiguous, *i.e.* if the same code can never be generated from more than one structure. That is the case for some of them, but not all. If a code is ambiguous, the structure you find might or might not be identical to the query. In terms of getting fast searching that may not matter as long as there is a second phase in which the retrieved structures are checked to find out which one, if any, is actually identical to the query. As mentioned in Chapter 6.1, comparing the connection table for a query with tables for tens of thousands of structures in a database would be too slow, but it is not a problem to compare it with the tables for just a handful of structures.

One way of comparing structures in detail is to use a backtracking algorithm.[17] Backtracking algorithms come in various forms, but here is a description of one. It assumes that you are comparing single, connected structures and would not work without modification for things like salts or mixtures because they contain fragments that have no formal connecting bonds between them. Also, it ignores stereochemistry.

If at any stage in the following procedure you find you have visited and matched every atom in both structures, then the structures are identical and you can terminate the comparison. Choose an atom in one structure – let us make it the structure of the query. Find an atom of the same type in the other structure, which we will call the target. Choose one of the bonds on the atom in the query and look for a bond of the same type in the target. If there is one, select the atoms at the other end of the bond in both structures and check that they are identical. If not, look for and try growing along another matching bond on the originally matched atom in the target. When you find a match, choose a bond on the second atom in the query, other than the one you just grew out along, and apply the same procedure to find the next matching bond and atom in the target. If at any point you run out of atoms or bonds, or fail to find a match, step back to the previous atom and try growing along a bond that you have not tried yet. You need to keep track of the atoms you have visited. If you come across one again while moving forward along a growing branch, you have found a ring and you must not traverse onto that atom. If you do not include this rule you will go round and round the ring indefinitely. If you run out of options for continuing to grow through the structures before you have visited every atom in them, then you mapped the wrong first atom in the candidate to the first atom in the query.

Start again, choosing a different matching atom in the candidate. Keep going until either you find a complete match or you run out of atoms to start from.

There are ways of minimising the amount of work you have to do in the above procedure. Trivially, if the numbers of atoms of each elemental type differ in the two structures they obviously cannot match and there is no point in going any further. Ullmann[18] suggested the following way to rule out a match in many cases without having to get into backtracking. Set up a matrix in which the rows represent the atoms in the query structure and the columns represent the atoms in the target structure. Set the value of each entry in the matrix to 1 if the attributes of the atom in the query structure match the attributes of the atom in the target structure, and otherwise to 0 (what attributes you use will depend on the design of your particular system – it might simply be the atomic number, for example). Now look at each matching pair of atoms, Q_i from the query and T_i from the target (*i.e.* the pairings for which you have put a 1 into the matrix). If none of the pairings of atoms attached to Q_i with atoms attached to T_i can match (*i.e.* there is a 0 in the matrix for all of them) then clearly you will not be able to construct a mapping for the structure that grows from the pairing of Q_i with T_i. So change the 1 in the matrix for the pairing (Q_i, T_i) to 0. At the end of this process, if any row in the matrix is all zeros, that means there is no atom in the target that can be mapped successfully onto the atom in the query associated with the row. That being the case the structures cannot be mapped onto each other – they cannot be the same.

Once you do get into backtracking, you can minimise the potential number of unsuccessful matches by starting the search from the least common type of atom in the query. Take 1,3-dibromohexane as an example. If you start from a carbon atom, you might make up to five wrong choices before hitting on the right one. If you start from bromine there are only two choices available. You are not limited to making use of what element each atom is – you can use whatever attributes of atoms are recorded in the connection tables.

6.3   Substructure Searching

A backtracking algorithm of the kind described in Chapter 6.2, works also for substructure searching, if you make the substructure the primary one and try to grow identical paths in the whole structure. The Ullmann algorithm works as long as you base the analysis of the matrix on the rows for the query as described in Chapter 6.2 (there will be columns for the target in the matrix where there are no matches, since the target is bigger than the query), and – rather obviously – provided that you are looking for maps that cover more than a single atom (since the Ulmann algorithm tests whether you can map both an atom and one of its neighbours). The completion point for the backtracking algorithm is having visited every atom in the query – you will not, of course, have visited every atom in the whole structure – but you need to apply the algorithm repeatedly until you have run out of places to start from in the whole structure because the substructure might be embedded in it more than once.

Whatever method you choose for substructure searching, how can you avoid having to apply it to every structure in a database? Unique coding systems to support fast searching, as described in Chapter 6.1, are no use because they are unique to the complete structure or substructure they describe. Obviously, a substructure query and the full structures onto which it can be mapped will not have the same code. The long standing solution to this problem has been to include "search keys" in the database. A set of substructural fragments is decided upon. They might include, for example, a list of individual atom types, fragments such as a carbonyl group or a phenolic hydroxyl group, a saturated six-membered ring containing only carbon atoms, one containing a heteroatom, various aromatic rings, and so on. Each fragment is assigned a specific bit position in a lengthy binary number often referred to as a "fingerprint". If a fragment is present in a structure then its bit is set to 1 in the fingerprint for the structure, otherwise it is set to 0.

At search time the fingerprint for the substructural query is generated and then it is compared bitwise using AND NOT logic with the fingerprints in the database. If the resultant number is zero then a potential match has been found: if a fragment is present in the query but not in the target structure (which means a match is impossible), the output at that bit position with be 1; if a fragment is present in the query and in the target structure (the sought for situation) the output will be 0; if a fragment is absent in the query the output will be zero whether the fragment is present in the target or not (which is the behaviour we want because its presence in the target is irrelevant).

With judicious choice of keying fragments, this method reduces the number of candidates for atom by atom matching to a handful for all but the most challenging substructure queries and it allows very fast substructure searching of databases containing hundreds of thousands of structures. It is easy to think of substructure queries for which search keys will achieve very little, for example searching for all structures containing a carbon atom in a database of organic structures. In practice, it is unusual for a substructure search to be so vague. Simply returning and reporting such a huge answer set would take a long time anyway, and the user would expect to have a long wait. The art of designing a set of search keys is to choose keys that are maximally effective in reducing the typical number of hits returned for atom by atom matching. Statistically, an ideal search key is one that is present in half the structures in a database, but that has to be balanced against the need to have an even distribution of keys. Obviously, having several keys in the same 50% of the structures is less useful than would be the same number of keys with a better spread.

Search keys, since they are available in databases, get used for other purposes as well as to support substructure searching. They have been used as fragments for statistical learning algorithms in systems to predict pharmacological and toxicological activity, a topic covered in Chapter 11. They are used in some database systems to support similarity searching, although in others a separate set of fragment descriptors is used, the first being tailored to efficient substructure search keying and the second to similarity searching. The idea behind using keys of this kind for similarity searching is that the more keys two structures have in common, the more similar they must be. Note that this is in the context of structural similarity. Similarity is context dependent. Potatoes and yams are similar if you are interested in classifying foods according to their nutritional function: they are dissimilar if you are classifying crops according to their suitability for growing in northern Europe. The issues of similarity, how to define it, and how to measure it are the subjects of a substantial field of research that is not covered in this book.

It is convenient to mention here that "circular fingerprints" are used in many applications. They are defined by an atom and its environment. For example, bits may be set according to the element type of the atom, according to the types of elements attached alpha to the atom of current interest, and those attached beta to it, and so on. For some purposes it is sufficient just to describe out to the first shell around the central atom (*i.e.* to include the central atom and those alpha to it) but in some applications much larger shells are included. If the shells are big enough, each atom carries with it something approaching a description of its environment in terms of the whole molecule and it may be unique even in a large set of structures. That might appear useful, but it does not help if you are looking for a substructure, since the fingerprints for the substructure obviously will not match with any in the database that encompass a surrounding structure bigger than the query. So the decisions to use circular fingerprints and how large to make them, depend on the purpose for which they are to be used.

6.4   Set Reduction

Set reduction, in which you progressively eliminate atoms that cannot match in a pair or collection of structures, can greatly reduce the need to backtrack through structures.[19,20] The aim is to find atoms in one structure that do not have counterparts in the other structure and eliminate them from consideration. The Ullmann algorithm, described in Chapter 6.2, is an example and so is the use of fingerprints described in Chapter 6.3. One could apply the Ullmann algorithm, or a similar algorithm, with elemental type as the characteristic to be compared between atoms. For example, if one structure contains a sulfur atom and the other contains no sulfur there is obviously no point in looking for overlaps between the structures that require mapping to the sulfur atom. Just comparing the elements present in each structure might be a start but it would not allow you to eliminate many atoms from a typical pair of organic structures. In practice, more distinguishing attributes, or sets of attributes, are used. An example is the EXTENDED CONNECTIVITY value in the Morgan algorithm that describes the environment of an atom in a graph (see Chapter 6.1.2) and is thus rather like a circular fingerprint in its information content but differently expressed.

Peter Johnson and Chris Marshall[20] were interested in finding matching parts of the carbon skeletons of pairs of structures, initially without regard to functionality. They use set reduction based on the locations of atoms in a graph, but they compute values in a different way from the one used by Morgan to calculate EXTENDED CONNECTIVITY. At a first level, they label each atom with the number of carbon atoms adjacent to it. To calculate the values for a second level they multiply the existing label for an atom by five and then add to it the sum of the squares of the labels of adjacent atoms. There is nothing deeply significant about this method of calculation. It is just one that causes values to diverge rapidly and minimises the frequency with which atoms in different environments get the same label coincidentally. The examples Johnson and Marshall use in their paper are shown in Figure 6.3 and the labels at level 1 and level 2 are shown in Figures 6.4 and 6.5. New levels are generated iteratively until a level is reached at which: there are no changes to adjacent atoms for any atom in the structures, which happens if the complete skeletons of the two structures are identical; or a level is reached in which only one pair of matching atoms remains; or a level is reached at which there remains no set with members in both structures, in which case the previous level is designated as the last one.

Table 6.3 shows the matching sets of atoms at each level for Structures 6.4 and 6.5. At level 4 only one matching pair remains, 'a' and 'm', and they must mark the limit of a matching region. There is only one atom alpha to atom 'a' and one to atom 'm' – atoms 'b' and 'n', respectively – and so we can assume that they should be mapped onto each other. Similarly we can map atoms 'c' and 'o'. Moving further along the growing fragment, there are two atoms adjacent to atom 'c' – atoms 'd' and 'g' – and two adjacent to atom 'o' – atoms 'p' and 'q'. If you look at level 1 in Table 6.3 you will see that 'd' is in the same set as 'q' and 'g' is in the same set as 'p', and so we can map them accordingly.

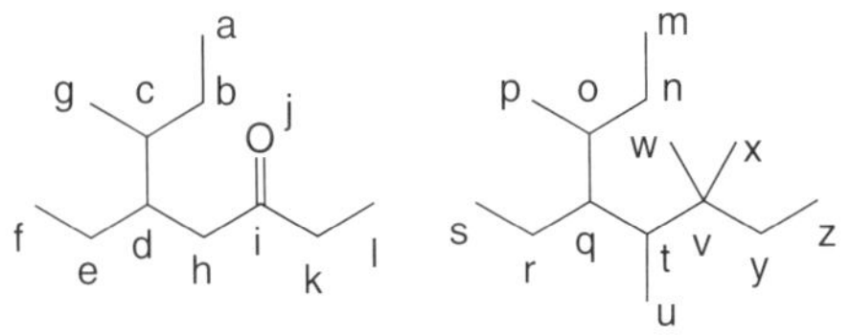

Figure 6.3 A pair of structures for mapping to find what they have in common.

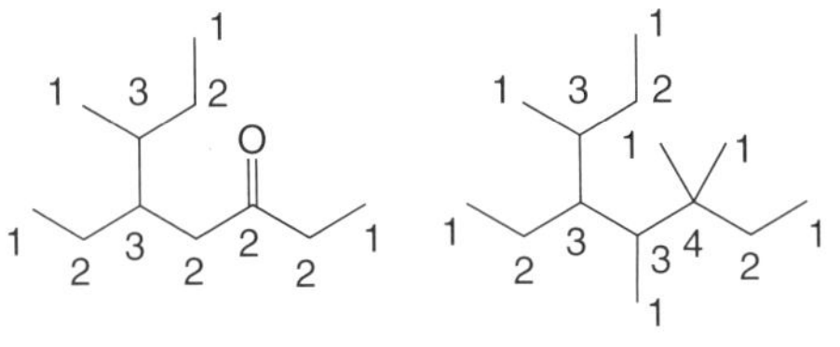

Figure 6.4 Atoms labelled for level 1 according to the number of carbon atoms attached to them.

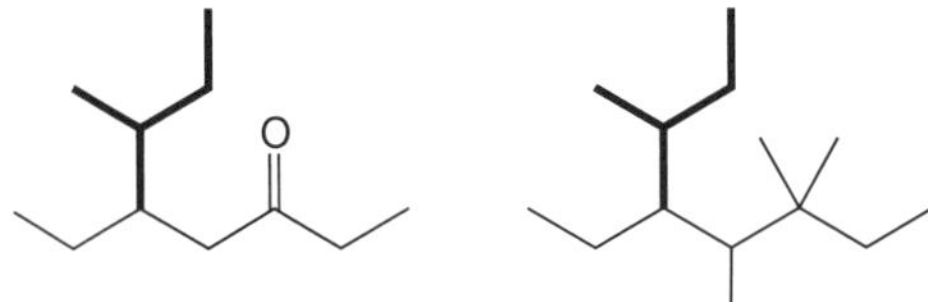

Figure 6.5 Atoms with labels calculated for level 2 from those in Figure 6.4.

Table 6.3 Sets of atoms with the same computed labels in Structures 6.4 and 6.5 at each level.

Level	Set number	Atoms in Structure 6.4	Atoms in Structure 6.5
1	1	a, f, g, l	m, p, s, u, w, x, z
	2	b, e, h, i, k	n, r, y
	3	c, d	o, q, t
2	1	a, f, l	m, s, z
	2	g	p, u
	3	b, e	n, r
	4	c	o
3	1	a, f	m, s
	2	b	n
	3	g	p
4	1	a	m

Figure 6.6 The part map formed by growing from atoms 'a' and 'm'.

We have found the mapping shown with heavy lines in Figure 6.6, described by Johnson and Marshall as a part-map for reasons that will become apparent.

Now look at the sets at level 3. We have already discovered the pairings of 'g' with 'p', 'b' with 'n', and 'a' with 'm', but the pairings of 'f' with 's', 'a' with 's', and 'f' with 'm' are new. If you follow each of these leads in the way that we followed the one that paired 'a' with 'm', you end up with the three new part-maps in Figure 6.7. Starting again from the sets at level 2 eventually leads to a set of part-maps of two atom/one bond fragments which I will not list here, although I shall come back to the one shown in Figure 6.8. The part-maps in Figures 6.6 and 6.7(a) share a common atom – atom 'd' in Structure 6.4 is paired with atom 'q' in Structure 6.5 in both part-maps – and so the part-maps

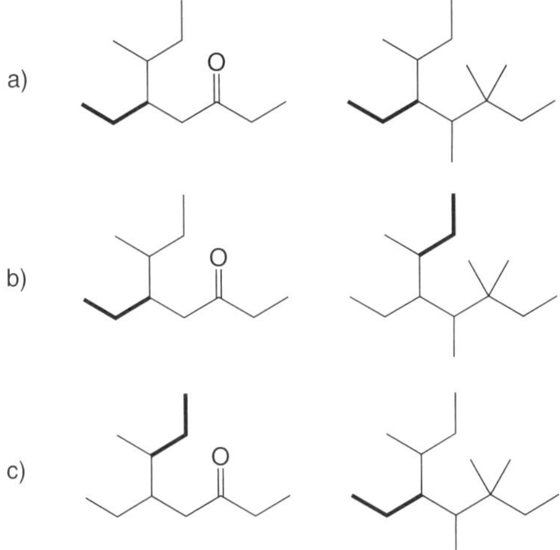

Figure 6.7   New part-maps found by studying the sets of atoms at level 3.

Figure 6.8   One of six further mappings that can be found by studying the sets of atoms at level 2.

Figure 6.9   Valid combination of the part-maps shown in Figures 6.6, 6.7 and 6.8.

can be combined. There is no conflict between any of the atom mappings for this combined map and the part-map shown in Figure 6.8 and that can also be added to the combination. Now we have the part-map shown in Figure 6.9. The authors say that the shortest path is now grown between the mapped fragments of Figure 6.9 in each of the structures. If the resultant paths are of equal length, as they are in this case, the connection can be made, leading to the complete map shown in Figure 6.10. They do not say how they grow the connecting path. You might like to work out a suitable method.

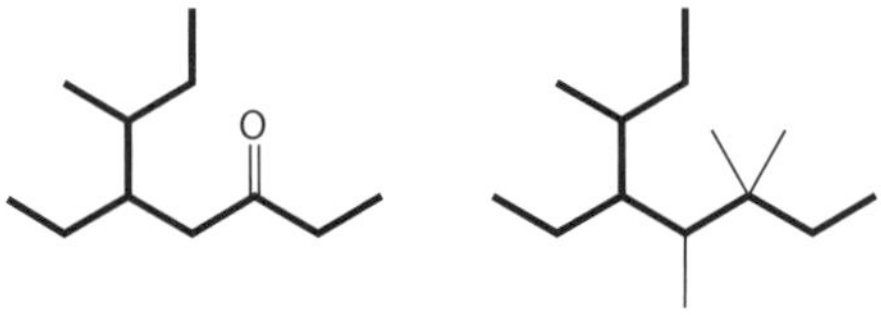

Figure 6.10   The completed map.

The authors go on to talk about building imperfect maps, in which one or more carbon atoms would have to be added or removed to convert one structure into the other. They also describe situations in which, using their method, mapping can only be completed by doing some backtracking, and situations in which maps cannot be found, or too many are found. The above description shows you how set reduction methods can nevertheless greatly reduce the amount of computation you have to do to find correspondences between structures.

6.5   Superstructure and Markush Structure Searching

When you do a substructure search, you want to find complete structures that incorporate your query. When you ask whether a query structure contains a toxicological alert, or toxicophore, the situation is reversed; you want to find substructures in a database that are contained in your query. This kind of search is sometimes called a superstructure search. The methods that are used for substructure searching can be applied to superstructure searching by swapping the roles of query and candidate structures, but criteria for the best search keys, for example, are different.

The ultimate challenge is to search a database of Markush structures for Markush queries – *i.e.* to search a database of imprecisely defined structures for imprecisely defined queries. The term "Markush structure", for structures in which various parts are represented by symbols such as "X", "R", "R1", and "R2" with footnotes listing alternative substituents that the symbols represent, comes from the name of the first person to use such a representation successfully in a US patent application, Eugene Markush, and computer-based patent information services offer ways of doing Markush searching. However, they all have limitations for good reasons. Matching Markush structures against one another may not seem that much of a problem at first, but the more you think about it the more complicated it gets. Dealing with simple variations at one atom site may not look so difficult – for example, finding the match between Structures 6.6 and 6.7 in Figure 6.11 – but try thinking about what would happen if Structures 6.6 and 6.7 each contained half a dozen variable components rather than one and the variable components were more complicated than X and Y. It is not unusual to find definitions in patents of the form "where R^1 represents hydrogen, hydroxy, alkyl, halo, haloalkyl, hydroxyalkyl, alkenyl,

6.6 **6.7**

X = O or S   Y = O or NR

where R = H or methyl

Figure 6.11 Markush structures that partly overlap.

alkynyl, cycloalkyl, aryl, aralkyl, heterocyclyl or a group -S(O)nR^2 where $n = 0$, 1, or 2, and R^2 represents an alkyl, aryl or aralkyl group or, when $n = 0$, a hydrogen atom."[21] Markush searching being an area of great commercial interest, relatively little has been published about it and the publications that there are tend to be about the strengths and weaknesses of commercial online search services rather than about the technicalities of how they work. I have included one or two references at the end of this chapter.[22-25] If you can crack the problem of searching Markush databases for Markush queries exhaustively, precisely, and fast you are on your way to your first million euros or dollars (probably both).

6.6 Reaction Searching

Pretty well all of what has been said earlier in this chapter about structure and substructure searching applies also to reaction searching, but there is an important additional requirement. Searching for a reaction in a general way, as distinct from searching for a specific instance of it – for example, searching for all the Diels–Alder reactions in a database as distinct from searching for the formation of dicyclopentadiene from cyclopentadiene – is the reaction-searching equivalent of a substructure search. A reaction is defined by the changes that take place at the reaction centre. However, it is not sufficient to define a reaction only in terms of the differences between the starting materials and products. If you do not map atoms from one side of a reaction to the other you leave out important information. Either the hydroxyl oxygen atom of the alcohol or that of the acid really does map specifically to the oxygen atom in the

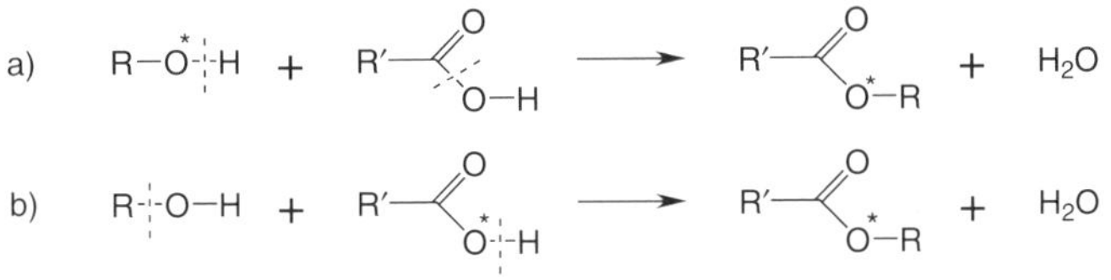

Figure 6.12 Formation of an ester with retention of the oxygen from the alcohol or the acid.

product ester in Figure 6.12, and there is no justification for being casual about it. Not all reactions stored in reaction database systems have the correct mapping and if the mapping is wrong, searches for reactions with mechanisms corresponding to Figure 6.12(a) (the more usual but not the only one) or Figure 6.12(b) will give wrong answers. Errors will become apparent if one of your oxygen atoms is the ^{18}O isotope, for example.

References

1. D. Weininger, SMILES 3. Depict. Graphical Depiction of Chemical Structures, *J. Chem, Inf. Comput. Sci.*, 1990, **30**, 237–243.
2. M. Bersohn, A Sum Algorithm for Numbering the Atoms of Molecules, *Comput. Chem.*, 1978, **2**, 113–116.
3. G. Neglur, R. L. Grossman and B. Liu, in *Proceedings: Data Integration in the Life Sciences, 2nd Int. Workshop, DILS 2005, San Diego, CA, USA, July 20–22, 2005*, ed. B. Ludäscher and L. Raschid, Springer, 2005, pp. 145–157.
4. H. L. Morgan, The Generation of a Unique Machine Description for Chemical Structures – a Technique Developed at Chemical Abstracts Service, *J. Chem. Doc.*, 1965, **5**, 107–113.
5. W. T. Wipke and T. M. Dyott, Stereochemically Unique Naming Algorithm, *J. Am. Chem. Soc.*, 1974, **96**, 4834–4842.
6. *Encyclopedia of Computational Chemistry*, ed. P. von Rague, Wiley, Chichester UK, 1998, pp.167–168.
7. J. Braun, R. Gugisch, A. Kerber, R. Laue, M. Meringer and C. Rücker, MOLGEN-CID, a Canonicaliser for Molecules and Graphs Accessible Through the Internet, *J. Chem. Inf. Comput. Sci.*, 2004, **44**, 542–548.
8. J. B. Hendrickson and A. G. Toczko, Unique Numbering and Cataloguing of Molecular Structures, *J. Chem. Inf. Comput. Sci.*, 1983, **23**, 171–177.
9. M. Randić, G. M. Brissey and C. L. Wilkins, Computer Perception of Topological Symmetry via Canonical Numbering of Atoms, *J. Chem. Inf. Comput. Sci.*, 1981, **21**, 52–59(and a series of earlier papers cited in this one).
10. S. E. Stein, S. R. Heller and D. Tchekhovskoi, in *Proceedings of the 2003 International Chemical Information Conference (Nimes)*, Infonortics, pp. 131–143.
11. ChemDraw is supplied by CambridgeSoft Corporation, Cambridge, Massachusetts, USA. http:// www.cambridgesoft.com
12. Marvin is supplied by ChemAxon Kft., Máramaros köz 3/a, Budapest, 1037 Hungary.
13. http://www.acdlabs.com
14. http://pubchem.ncbi.nlm.nih.gov
15. http://www.chemspider.com
16. N. Day developed the website at the Unilever Centre, Dept.Chem., University of Cambridge, England. http://wwmm.ch.cam.ac.uk/inchifaq/

17. L. C. Ray and R. A. Kirsch, Finding Chemical Records by Digital computers, *Science*, 1957, **126**, 814–819.
18. J. R. Ullmann, An Algorithm for Subgraph Isomorphism, *J. Assoc. Comput. Mach.*, 1976, **23**, 31–42.
19. John Figueras, Automorphism and Equivalence Classes, *J. Chem. Inf. Comput. Sci.*, 1992, **32**, 153–157.
20. A. P. Johnson and C. Marshall, Starting Material Oriented Retrosynthetic Analysis in the LHASA Program. 2. Mapping the SM and Target Structures, *J. Chem. Inf. Comput. Sci.*, 1992, **32**, 418–425.
21. A. Percival and P. N. Judson, *Br. Pat.* 1567781, *Appl.*, 20147/76, Filed 2nd May 1977.
22. A. von Scholley, A Relaxation Algorithm for Generic Chemical Structure Screening, *J. Chem. Inf. Comput. Sci.*, 1984, **24**, 235–241.
23. W. Fisanick, Storage and Retrieval of Generic Chemical Structure Representations, US Pat. 4 642 762, 1987.
24. J. M. Barnard, A Comparison of Different Approaches to Markush Structure Handling, *J. Chem. Inf. Comput. Sci.*, 1991, **31**, 64–68.
25. A. H. Berks, Current State of the Art of Markush Topological Search Systems, in *Handbook of Chemoinformatics: from Data to Knowledge*, Wiley VCH, Weinheim, 2003, **2**, pp. 885–903.

CHAPTER 7

Protons that Come and Go

7.1 Dealing with Tautomerism

It was mentioned in Chapter 3.2 that EROS treats hydrogen atoms on either side
of a reaction as being the same atom for pragmatic reasons while, in reality,
according to the accepted reaction mechanism, they often are not. The nomadic
propensity of protons presents all sorts of headaches for chemists and biologists,
as well as computer scientists, not least of which is deciding how to depict
structures that are capable of tautomerism. Learning about keto–enol tautomerism comes early in a chemist's training, with the aldol reaction (Figure 7.1)
providing an introduction to the importance of tautomerism to reaction
mechanisms. For simple ketones and aldehydes the equilibrium lies heavily in
favour of the ketone tautomer and convention has it that ketones are drawn as
such in chemical diagrams. When it comes to *beta*-diketones things are less clear.
Depending on circumstance, a chemist may draw a diketone, as in Figure 7.2(a),
or a keto–enol, as in Figures 7.2(b) and 7.2(c), or something vaguely in between,
as in Figures 7.2(d) and 7.2(e). Which picture the chemist chooses may depend
on where he or she wants to take us when presenting a mechanistic argument
about a reaction, may depend on assumptions about conjugation or other
interactions with nearby functionality in a bigger structure, or may be arbitrary.
Whichever depiction is chosen, other chemists realise that all three structural
forms, (a), (b) and (c), contribute to the properties and reactivity of the compound. Non-chemists, even if they know about tautomerism as a concept, will
not necessarily think of all its implications in the way a chemist should. If they
are looking for information about a compound in a book they may miss
instances where the author has presented the compound as a different tautomer;
so might a chemist unless he or she gave the matter some thought before starting
to leaf through the book: the eye can easily pass over a picture of a different

RSC Theoretical and Computational Chemistry Series No. 1
Knowledge-based Expert Systems in Chemistry: Not Counting on Computers
By Philip Judson
© Philip Judson 2009
Published by the Royal Society of Chemistry, www.rsc.org

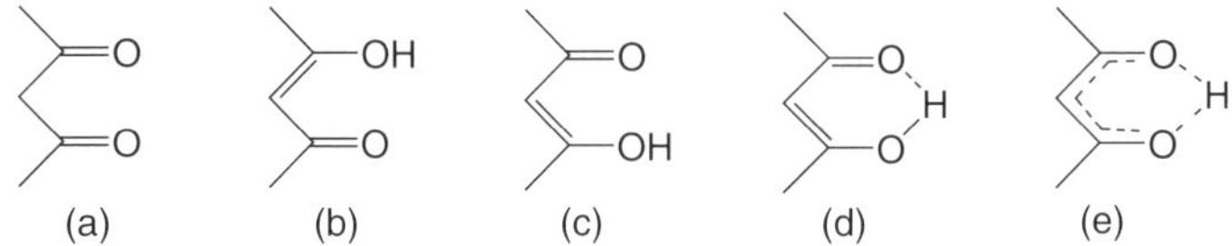

Figure 7.1   The aldol reaction.

Figure 7.2   Beta diketone tautomerism.

tautomer, and names in an index can be very different for tautomers of the same compound.

The diagram in Figure 7.2(e) might look like the answer to the problem but it is not. There is the practical point from a chemist's point of view that it is not suited to thinking about reaction mechanisms, or potential mechanisms, by "arrow pushing". More problematic is that while diagram (e) might be an acceptable compromise for this simple diketone (if it is all right to call it a diketone) the approach does not generalise well. Should all ketones be represented as somewhere between ketones and enols, even though they are predominantly in the ketone form? If so, which way should the enolisation go if the alkyl groups on each side of the ketone function differ? If we were to adopt diagrams like the one in Figure 7.2(e) as standard representations of tautomeric structures, how would we name them?

The preferred convention is to draw and name a compound as whichever tautomer is believed to predominate or suits the context in which it is being discussed. There are no difficulties with recognising diagrams, or finding entries in book indexes, for simple cases like ketones that are subject to accepted conventions, but for less obvious cases the chemist needs to consider the alternatives and look for all of them. 2-Pyridone, for example, is just as likely to be depicted and indexed as 2-hydroxypyridine, and you will miss many references if you search for only one of them unless the person who compiled the index anticipated the problem and included cross references.

It is obvious that the same problems will apply to storing chemical structures capable of tautomerism in a computer database. There are database systems in which each structure is stored in whatever tautomeric form took the fancy of the person entering it and end-users are expected to search for all possible options, but most systems are friendlier than that. Different systems use different solutions. Each has its advantages but none of them fixes everything.

One solution is to establish standard conventions for deciding on the single, "correct" tautomeric form for a structure and to impose them on the people who enter data into a system. As long as an end-user also obeys the conventions, retrieval should be reliable. Database staff and end-users need not necessarily learn and obey the rules. They can be built into algorithms in the software so that when anyone draws a structure it is automatically redrawn to conform to the conventions. But just ensuring consistency between structures stored in the database and entered as queries is not really enough. It matters whether the chosen tautomers are the ones most chemists would expect to see, because a chemical structure database system does not exist for the benefit of a community of computers; it is intended to be a service to human users and it should be suited to them. How do you come up with conventions to cope with all the complications? Starting with isolated ketones you might, sensibly, decide that the tautomer-checking software will convert all enols into ketones. Moving on to *beta*-diketones you might decide to stay with your convention. What about phenols? No-one is going to be happy with having them redrawn as ketones. So you introduce new rules to say that aromatic hydroxy compounds should not be redrawn and that ketones should be redrawn as enols if so doing converts then into aromatic hydroxy compounds. It is at that point that somebody reminds you about the pyridones and you realise things are not going to be easy.

A safer solution, used in some systems, is to generate all possible tautomers of a query when the user hits the "search" button and to look for all of them in the database or knowledge base. It does not matter which tautomers were drawn by the people entering the data. They will all be found. The user can be given the option to search exclusively for the query structure as drawn or to look for tautomers in addition. As well as being useful when the user wants to find references to a specific tautomer, giving the user this option speeds up searches when the user is confident that potential alternatives to the one drawn are unlikely to be of interest. Some structures have very large numbers of potential tautomeric forms and searching for all of them can take a long time. I was helping out some years ago at the exhibition stand of a company in direct competition with what was then Molecular Design Limited (MDL). We had recently stumbled upon a structure with dozens of tautomeric forms that could keep a search engine of the day busy for minutes if not hours. There was a new member of staff with us at the stand, not yet known to our competitors. We sent her along to the MDL stand where she waited until a good crowd of prospects had gathered and then asked if she could try searching for a structure she happened to have with her. It kept the software busy for longer than usual but not much: the MDL software team had had the sense to realise that rogue structures were bound to come up, even if not delivered by jokers from competitors' stands, and there was a limit on how far the program would go with generating and searching for tautomers before cutting its losses and just presenting the ones reasonably close to the query. Programs that use this approach usually limit tautomer generation to the more obvious cases. It keeps things simple and fast, and it is enough for practical purposes.

You can solve the problem from the other end, of course, and some systems do that: you can store all the tautomers of a structure in the database or knowledge base – or at least the main ones if the number of possibilities is too large. Whatever tautomer the end user draws as a query, the corresponding entry will be found. It is then the responsibility of the person entering data to decide what tautomers to include and, if need be, you can provide a tool to generate a list of them automatically from which he or she can choose.

Some systems support a combination of both approaches. Derek for Windows[1] is an example. Knowledge base writers can specify any number of tautomeric patterns in an alert (a set of substructural features associated with toxicity) in the knowledge base. Knowledge base writers enter just the patterns that are relevant to the model they are associated with, which often is only a single tautomer and rarely very many. At run time a set of obvious tautomers is generated automatically for the query entered by an end-user, based on some simple rules. The knowledge base is searched for patterns that match any of the tautomers in the set generated from the query.

Finally, some systems use an internal representation akin to the one in Figure 7.2(e). Structures displayed to users are specific tautomers, but when a structure is added to the database an algorithm creates a generalised tautomeric representation to be stored with it. When a query is entered, the same generalised representation is created from the query and matched against the tautomeric entries in the database. This is a powerful method and potentially fast, since it avoids the need for multiple, parallel searches, but developing a reliable algorithm is not easy and, as with the other approaches, there will be difficult cases that have to be excepted or only partially dealt with.

7.2   Implicit and Explicit Hydrogen Atoms

A different problem with hydrogen atoms started out by being peculiar to chemical information systems. It ought to have remained so, but instead it has extended into chemistry more widely. Particularly in the early days of chemical computation, computer memory size and disk space were major issues. A protein contains thousands of atoms. Even the sort of compound you might synthesise in an undergraduate laboratory experiment can contain anything up to sixty atoms. A computer system needs to store information about all of them and their associated bonds, which takes space, and to carry out computations atom by atom, which takes time. Software developers needed to find ways to get the numbers down and there is a fairly obvious one.

Consider a simple hydrocarbon. If it is saturated and acyclic it contains just over twice as many hydrogen atoms as carbon atoms. Given that carbon has a valency of four (normally), any valency unsatisfied by unsaturation can be assumed to be satisfied by bonds to hydrogen atoms. Chemists know this and do not normally waste ink drawing all the hydrogen atoms and the bonds to them in chemical diagrams. Actually, they do not label the carbon atoms either – convention has it that if there is an unlabelled node in a chemical graph it represents a carbon atom, as in Figure 7.3(b).

Organic chemical structures are similarly abbreviated in most computer systems. A structure is represented as a set of non-hydrogen atoms and a set of

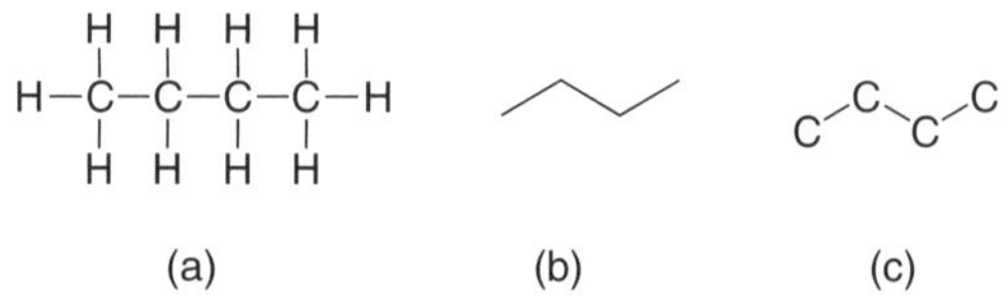

(a)   (b)   (c)

Figure 7.3   Representations of *n*-butane.

bonds connecting them. Each atom has attributes associated with it – the most obvious one being what element it is and another one being what isotope. The attributes of each bond include the bond order. Either the attributes of each atom include the identities of the bonds attached to it or the attributes of each bond include the identities of the atoms at its ends. So the number of hydrogen atoms attached to a carbon atom, for example, can easily be computed and there is no need to include the hydrogen atoms explicitly in the description of the structure. If a carbon has just one, single bond connected to it, it must also carry three hydrogen atoms. If it has just two single bonds or just one double bond connected to it, it must carry two hydrogen atoms, and so on. The hydrogen atoms are said to be implicit. To represent, say, *n*-butane with all its hydrogen atoms included would require the computer to handle fourteen atoms and thirteen bonds, as shown in Figure 7.3(a). Abbreviating the representation by leaving out the hydrogen atoms and the bonds to them reduces the requirement to four atoms and three bonds, as in Figure 7.3(c) (there is nothing to be gained in dropping the atom type from the computer representation since you have to assign the required memory space anyway, in case the atom is not carbon).

So far so good. The problems appear when you also treat hydrogen atoms attached to non-carbon atoms as implicit and when you do so in communications with human users. Historically, chemists always showed hydrogen atoms attached to hetero atoms in structural diagrams, implicit hydrogen substitution being assumed only for carbon atoms. It does not make sense for the greater part of the periodic table to assume that apparently unsatisfied valencies signify implicit hydrogen substitution and even omitting hydrogen substituents from first row elements in organic compounds can cause confusion.

Software developers tend to see things in terms of coding issues more than in terms of the end-users their systems communicate with and in consequence early computer programs omitted hydrogen atoms attached to hetero atoms as well as carbon atoms even in structures displayed to users. Chemical database systems quickly became widely used and their advent made chemical information and structure searching easily available for the first time to non-chemists. The omission of hetero hydrogen atoms, which had irritated chemists, was assumed by non-chemists to be correct convention and for them it is increasingly becoming so.

For elements beyond the first row of the periodic table, what oxidation state should be assumed if all hydrogen atoms are implicit? Indeed, how can a user specify an oxidation state? The established convention of making all hydrogen atoms on hetero atoms explicit avoids this difficulty. To take an extreme example, "P" represents elemental phosphorus and not PH_3 or PH_5. "Me_2S" represents what it appears to represent, and not the admittedly rather less likely Me_2SH_2. Still thinking about the problem in terms of their algorithms and internal representations of structures, some software developers added "lone pair" as a kind of pseudo-atom. A user not wanting the program to compensate for unsatisfied valency by adding hydrogen atoms could, so to speak, block the space by attaching one or more lone pairs of electrons. It is a clumsy solution – a kind of logical equivalent to a double negative – unsatisfactory for a chemist and mystifying for a non-chemist. Attaching charge has a similar effect. So, for example, "RO" is interpreted by most computer programs as "ROH" but "RO^-" is recognised as an alkoxide ion. A chemist needs to be hawk-eyed to notice that lurking in the heart of a complicated structural diagram there is "–O" where "$–O^-$" was expected, or vice versa. A non-chemist is unlikely even to think that it matters. And the problem has spread beyond the world of computing, with scientists frequently now omitting hydrogen atoms from hetero atoms in the diagrams that they draw.

It is at this point that the wonder of compromise steps in to maximise confusion. Modifying the programs to revert to established convention – *i.e.* always to show hydrogen atoms attached to hetero atoms explicitly – would be sure to upset users who have become accustomed to things the way they are, even if the computer and information scientists could be persuaded to do it. Instead, most programs allow users to choose whether hydrogen atoms should or should not be displayed on hetero atoms. It is a compromise we probably have to live with but for the worrying reason that half of the user community does not understand why it is an issue. And it is not really a solution . . .

It was mentioned earlier in this chapter that, rather than being treated as atoms in their own right, hydrogen atoms are treated as attributes of the atoms to which they are attached. So, in terms of representation within the computer, methane is not a carbon atom, four hydrogen atoms, and four single bonds: it is a carbon atom which has the attribute of carrying four hydrogen atoms. The notion of hydrogen atoms as attributes of the atoms to which they are attached is often expressed in computer displays by using a so-called super-atom representation. For example, ignoring the rest of the molecule, a secondary amine group is represented in displays as in Figure 7.4(a) without the implicit hydrogen displayed and as in Figure 7.4(b) with it displayed. Do not, however, assume that the diagrams shown in Figures 7.4(b) and 7.4(c) mean the same thing. Figure 7.4(b) represents an atom called "NH" with an elemental attribute = "N" and a hydrogen count attribute = 1. Figure 7.4(c) represents an atom called "N" with elemental attribute = "N", an atom called "H" with elemental attribute = "H", and a bond connecting them with bond order attribute = "single". In the absence of code to treat "H" as a special case, routines in a computer program for dealing with implicit hydrogen atoms will see a total of three bonds to the nitrogen atom, and the nitrogen atom will be given a calculated hydrogen count attribute = 0.

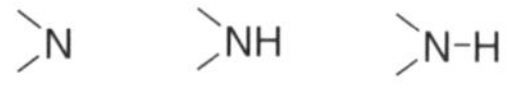

$$\overset{\diagdown}{\diagup}\text{N} \qquad \overset{\diagdown}{\diagup}\text{NH} \qquad \overset{\diagdown}{\diagup}\text{N-H}$$

(a)   (b)   (c)

Figure 7.4   Representations of secondary amines in computer displays.

I imagine you hope that the software developers have seen this coming and included code to sort out the confusion. Don't bank on it. Try experimenting with programs that you use. Don't forget to give them opportunities to trip each other up as well. For example, try using an application such as Isis Draw, ChemDraw, or Marvin as the graphical input for another application. Draw structures with and without hydrogen attached to a hetero atom, and try both drawing a connecting bond to an explicitly drawn hydrogen atom and using a super-atom representation (*e.g.* "OH", "NH", or "NH$_2$"). I promise you will get surprises unless you already knew about the problems.

I conducted a highly unscientific study last week. I wandered into an organic synthesis research laboratory in the chemistry department of Leeds University, found a couple of young researchers and asked each of them to draw phenol for me. I was reassured to find that they both still represent the phenol group as "–OH". I told them why I was interested and they both said that they used computer systems and knew that hydrogen atoms could or should be omitted when entering structures into most of them. They nevertheless felt that the correct way to draw chemical structures was to show hydrogen atoms on hetero atoms. A day or two later I was talking to a software developer in Germany. He went further, and said that the decision by early software developers not to show explicit hydrogen atoms on hetero atoms was downright incompetent and should never have been accepted. Like me (if you have not worked out my position by now) he was at a loss to understand why users have not made more of a fuss and nothing has been done about it after thirty years.

As you are reading this book you must have at least a passing interest in chemical information software. Chances are that you will one day be involved in designing or writing some. If you do nothing else for posterity, have your software put hydrogen atoms on hetero atoms and make it act sensibly with explicit and implicit representations (and indeed with the user of super-atoms in general).

Reference

1. Information taken from the help for Derek for Windows, version 10.0.2, supplied by Lhasa Limited, Leeds, England.

CHAPTER 8

Aromaticity and Stereochemistry

Aromaticity and stereochemistry present problems for computer systems – in both cases because of the uncertainties associated with them.

8.1 Aromaticity

Aromaticity has implications for the properties and reactivity of a compound. When you carry out a substructure or reaction search in a database, or enter the keying features of a structure for a reaction or retro-reaction into a knowledge base, you may want to specify that certain atoms or bonds must, or must not, be in an aromatic ring, albeit any kind of aromatic ring.

At different times in history, carbocyclic aromatic rings have been drawn in chemical diagrams just as simple hexagons – the chemist understanding by convention that they represented benzene rings and not cyclohexane rings – or as hexagons with solid or dotted rings inside them to symbolise the delocalisation of the electrons. Current practice is usually to draw localised, alternating, double and single bonds. A chemist understands that the rings are nevertheless aromatic. The chemist also knows that cyclo-octatetraene is not aromatic even though it is also represented as a ring of alternating double and single bonds. Many substituted heteroaromatic compounds are drawn differently by different chemists, or to suite the context in which a drawing is being used – for example, 2-pyridone or 2-hydroxypyridine. Whichever way the rings are drawn, a chemist realises that they, too, have aromatic character. A chemist knows, but how is a computer system to know?

Early computer database systems expected users to be specific about aromaticity, using a convention such as the one shown in Structure 8.1, and some systems still do. It has the advantage of being clear cut and straightforward to implement. The drawback is that the user must be aware of the convention and must get it right. A user who draws bonds in a ring in a substructure query as

RSC Theoretical and Computational Chemistry Series No. 1
Knowledge-based Expert Systems in Chemistry: Not Counting on Computers
By Philip Judson
© Philip Judson 2009
Published by the Royal Society of Chemistry, www.rsc.org

shown in Structure 8.2 will not find entries in the database in which the bonds have been recorded as aromatic ones. This may seem so obvious and simple that any chemist could be expected to use the right convention for the intended query. But it is less simple when it comes to something like 2-pyridone. Did the team who entered structures into the database treat 2-pyridones as aromatic compounds or not? What about less common heterocyclic systems, or cases where there is real debate about whether a ring is aromatic? Coming back even to clear cut cases, does a non-chemist who uses the database system have to learn about aromaticity in order to know how to draw queries? And what happens if nobody draws the query because it has been generated by some other piece of software?

8.1   8.2

Most chemical structure computer systems have some kind of automatic recognition of aromatic rings but conventions vary and there are some surprises for a user who is not forewarned. The most simplistic convention, and one still in use in some chemical database systems, is to treat all rings with alternating double and single bonds as aromatic and to treat everything else as non-aromatic. This is not a valid definition of aromaticity, of course, just a convention that is easy to implement and to explain to non-chemist users. It considers a compound like furan, Structure 8.3, to be non-aromatic and it considers cyclo-octatetraene to be aromatic. A substructure search for aromatic carboxylic acids, perhaps represented as shown in Structure 8.4, where C* indicates that the carbon atom must be in an aromatic ring, would find Structure 8.5 but not Structure 8.6, and it would find Structure 8.7 – not the results a chemist looking for aromatic carboxylic acids would want to see.

8.3   8.4

8.5   8.6

8.7   8.8

A minor improvement, and one used in some systems, is to limit aromaticity to six-membered rings containing alternating double and single bonds. That does eliminate the mis-classification of cyclo-octatetraene but it still leaves aromatic rings of other sizes mis-classified. A better approach is to define a more complete set of ring types that are considered aromatic, but the set is often surprisingly limited. For example, a typical definition for aromatic rings is those that are six membered with alternating double and single bonds and those that are five membered and contain two double bonds and one hetero atom. Benzene, furan, and 2-hydroxypyridine are recognised as aromatic, but 2-pyridone is not. If the system checks for tautomers, it may take account of the aromaticity of the 2-hydroxypyridine tautomer of 2-pyridone, depending on how tautomer processing is handled, but N-substituted pyridones will always be treated as non-aromatic.

"But why not use the Hückel rule?" I hear you ask. "If a ring contains $(4n + 2)$ π electrons it is aromatic". That is the solution used in some systems and it is a rule of thumb that provides a good approximation. There are popular examples of structures for which it fails, such as hexamethylene-cyclohexane, Structure 8.8, which has a ring containing six π electrons but is not aromatic, but they are unusual. The Hückel rule has the advantage that a computer system can apply it to any ring system, without the need for the ring to have been anticipated by the software designer or a rule- or data- base writer. Some programs combine the Hückel rule with a set of structure descriptions for exceptions, which would include hexamethylenecyclohexane. The program first applies the rule. If the structure is classified as aromatic, the program checks it against the set of exceptions and declassifies it if necessary.

Calculating electron densities in order to decide whether a ring is aromatic might seem the ultimate solution for a computer system, computers being good at doing arithmetic, but there are two problems. The main one is that the computational overhead may make processing too slow for convenience. There are fast methods for getting approximate values, but if the values are too approximate little might be gained over simply using the Hückel rule. A second problem, to which we will return shortly, is that even if you could do calculations with great precision you would find structures for which the answer was unclear because the reality is that aromaticity is not either "on" or "off" – there is a continuum between the aromatic and non-aromatic states. To use a chemists' favourite, is phloroglucinol (1,3,5-benzenetriol or 1,3,5-cyclohexatrione) aromatic? So at the end of all your calculations, you would still have to make arbitrary decisions in some cases.

The designers of the majority of systems have concluded that the best compromise is to use a specific set of patterns for rings that are considered to be aromatic. As mentioned a couple of paragraphs above, the sets of patterns that are used are surprisingly limited in some popular database management systems. Try doing substructure searches in different systems and see what comes out. You will still find systems that only recognise six membered rings with alternating double and single bonds as aromatic. Others will recognise well-known five-membered heterocyclic aromatic systems such as furan. They may or may not also recognise the cyclopentadienyl ring as aromatic, and they may not recognise even the text book seven-membered aromatic heterocycles as aromatic.

Part of the difficulty with handling aromaticity in computer systems arises from their insistance upon a firm decision: either a ring is aromatic or it is not. If you want to look for all structures containing a substructure in which a specified atom is, or is not, aromatic, a chemical database management system will give you a way to express your query, and within the limitations of its definition of aromaticity the system will find you answers. But what if you want to see structures in which the atom *might be* aromatic, rather than necessarily *is* aromatic? That is not the same as asking to see both the set of structures in which the atom is aromatic and the set in which it is not. Such a query would return cases where the atom was in a cyclohexane ring, or was even acyclic, depending on how you formulated your query. In neither of those cases would the atom fit the description of one that "might be aromatic". As this book progresses, problems to do with certainty – or rather uncertainty – and how to deal with them will feature increasingly.

8.2   Stereochemistry

Another area of structural uncertainty is stereochemistry. A given tetrahedral centre, for example, may have either one chirality or the other, or the chirality may be unknown. An important paper on representation in computer systems of tetrahedral stereochemistry and stereochemistry around double bonds was written by Todd Wipke and Thomas Dyott,[1] and most systems use the kind of approach they describe. Wipke and Dyott commented on the wide variety of conventions for dealing with stereochemistry in structural diagrams in books and journals, pretty well all of which had weaknesses. The situation thirty odd years later is probably better but there are still many cases to be found of ambiguous stereochemistry in diagrams in published papers and there is still no accepted standard for chemical drawings that covers all stereochemical possibilities.

8.2.1   Tetrahedral Centres

Starting with tetrahedral stereochemistry, consider Structure 8.9. Convention has it that the observer is to assume that 'c', 'd', and the bonds linking them are in the plane of the paper; 'a' is above the plane of the paper and 'b' is below it. Structure 8.10 might look different at first sight but it is the same as Structure 8.9. You cannot define the stereochemistry as *R* or *S* according to the Cahn–Ingold–Prelog rules because you do not have any information about the attachments 'a', 'b', 'c' and 'd'. You would be able to do so if you had a complete structural diagram, but a computer system needs to be able to deal with sub-structural fragments and Markush diagrams as well.

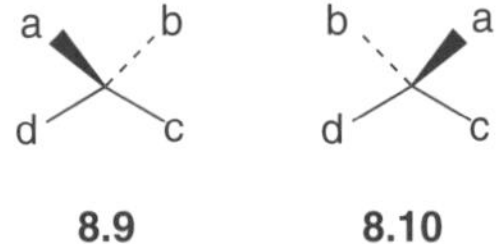

8.9   8.10

The simplest method for generating an ordered list of substituents that defines the stereochemistry is as follows. Imagine holding the three dimensional structure so that you are looking down the bond from atom 'a' towards the central atom. List the atoms, starting with 'a' followed by the other atoms in clockwise order. As mentioned in Chapter 6.1.2, there will be three possibilities – 'abcd' 'acdb' and 'adbc' – but any one of them allows you to reverse the process and redraw the correct stereoisomer. If you take this approach, the dotted bond to 'b' in the structural diagrams is surplus to requirements. Alternatively, and this should be no surprise, you can use the information that 'b' is below the plane to generate a similar set of lists if you look down the bond from the central atom towards 'b' and put atom 'b' last in the lists – 'acdb', 'cdab' and 'dacb'. One of the three lists is the same whichever method you use ('acdb'), which is reassuring. So the stereochemistry can be represented more succinctly in the 2D diagrams by showing only either the up bond or the down bond, as in Structures 8.11 and 8.12. If you do some mental acrobatics you will see that all the lists tell the same story, whether you assume looking down from the first atom or down towards the last, even though reverse engineering of the different lists will lead to different 2D pictures (*e.g.* Structure 8.13, created by making the first atom in 'cdab' an up atom although the list was based on making the last atom, 'b', a down atom).

8.11   **8.12**   **8.13**

Wipke and Dyott describe a more general method to deal with a variety of representations of stereochemistry that involve more than one dotted bond or wedge bond, but the end results are the same. They also cover the well-established conventions of omitting hydrogen atoms, as illustrated by Structure 8.14. The authors list a collection of representations that are either ambiguous or self-contradictory and which therefore cannot logically be interpreted and should not be used.

8.14

The safest way to avoid ambiguity is normally to choose only one bond and to mark it either as up or as down. Even so, there are some interesting traps for the unwary. For example, consider dicyclopentadiene, Structure 8.15. In your imagination you tilt the structure so that you are looking at it from somewhere below, and it appears as Structure 8.16 (the purpose of the numbering on four of the atoms will become apparent). Thinking you now know how to define the stereochemistry by marking the bonds to the bridge hydrogen atoms, you draw Structure 8.17a. But look at atoms 1, 2 and 3. Viewed from the angle shown in

Structure 8.16, with hydrogen atom 4 pointing away from you, you were looking at them from inside the cage and they were in clockwise order 1, 2, 3. In rocking the structure back to the way it is in Structure 8.17 you have moved your point of view. You now see the three carbon atoms in clockwise order 1, 3, 2 because you are not looking from inside the cage any more – you are looking from the outside, *i.e.* down onto it from hydrogen 4. So, to define the stereocentres by looking towards them from the "up" atom and listing the other substituents in clockwise order, the correct representation is Structure 8.17b. You must look locally at the stereocentre you want to define, imagine moving to a position where you see three of the atoms in the 3D structure in the same plane and still in the same clockwise order as in the 2D picture, and ask yourself whether the fourth atom points up or down.

8.15

8.16

8.17a

8.17b

Although I have ignored it in these diagrams, there is asymmetry at the ends of the fusion bond between the 6- and 5-membered rings in dicyclopentadiene. If you add the two hydrogen atoms and think about whether to mark each of them as "up" or "down" you will find that you are much less likely to make the error that arose in the case of the bridgehead hydrogen atom at the back of the molecule. The angle from which you happen to be viewing the 3D picture is roughly consistent with the angle from which you need to view each of these stereocentres to decide how to represent the stereochemistry in a 2D structural diagram.

Some chemical structure computer systems ignore tetrahedral stereochemistry but most of the major ones take it into account at least at carbon atoms. When you search in chemical structure databases, for example, you can usually elect to search for an exact enantiomer or for all its stereoisomers. Computer systems develop incrementally as needs arise in practice. In consequence most chemical structure systems are limited to dealing only with carbon and do not support searching for specific configurations around tetrahedral centres at heteroatoms.

Unfortunately for computer software designers, chemists use wedged bonds for two different purposes in structural diagrams, and only associated text makes it clear which is meant. Take Structure 8.14. Does it represent a single stereoisomer, or does it represent the pair in which the methyl group and bromine are up and down relative to each other? The convention for some computer user interfaces is that the diagram as shown indicates only relative stereochemistry, *i.e.* it represents the pair of isomers. If it indicated absolute stereochemistry it would be flagged as such. For example in earlier software originating from MDL Information Systems the word "chiral" appeared beside the structure. There are difficulties with this all or nothing approach if the absolute stereochemistry of one part of a structure is known but only relative stereochemistry is known elsewhere. A more advanced representation[2] that came from MDL gives much more complete coverage of the possibilities. Each stereocentre is individually marked. If it is marked "absolute" then the stereochemistry that is represented is absolute. Other centres can be grouped and classed "or" or "&". If a centre is flagged "or" it means that the substance is a single isomer but it is not known which one. If a centre is flagged "&" it means that the substance is a mixture of both isomers.

8.2.2 Double Bonds

Geometrical isomerism about double bonds is more easily dealt with because the substituents around a double bond are all in the same plane and their disposition can be set out easily in a 2D diagram. Nevertheless, double bonds cause headaches for software developers. Uncertainty about the stereochemistry at double bonds is surprisingly common. There are many cases where rotation can occur in mesomeric forms of a structure that is formally drawn as though it contained a stable double bond. Extreme examples that I came across when I worked in organic synthesis were so-called "push-pull alkenes" such as Structure 8.18[3] in which there is free rotation of the formal double bond at room temperature. Even where compounds contain an isolated double bond not given to this kind of behaviour, it is not always clear which isomer someone has written about.

8.18

MDL Information Systems were, I think, the first to introduce the "crossed bond" in commercial software as a simple convention for representing uncertainty about geometrical isomerism. In systems where the convention is used, Structures 8.19a and 8.19b represent specifically the isomers shown. The diagram used for Structure 8.19c indicates that either the stereochemistry is not known or both isomers are present. It is a simple and convenient representation with which you might think there should be no problems. The problems that do arise stem from ambiguities in printed scientific papers combined with concerns about presentation style in computer systems.

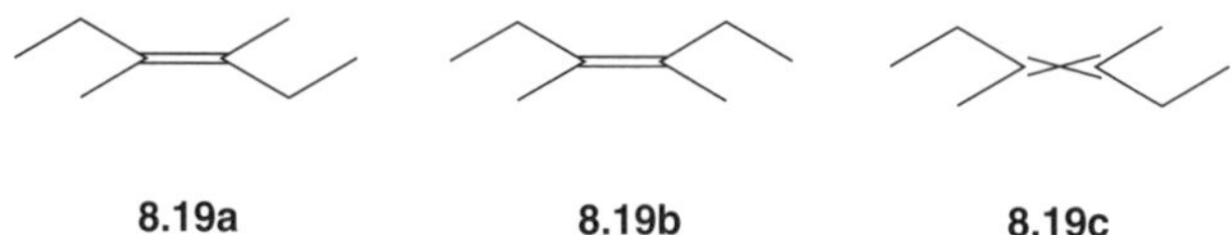

8.19a   **8.19b**   **8.19c**

As they were introduced for computer systems relatively recently, crossed bonds are not often found in hard-copy publications. An abstractor preparing data for a computer database has some difficulties as a result, and automated systems for recognising and transcribing graphics in chemical papers are worse placed. The diagrams in a typical paper will show double bonds as conventional double bonds. Whether the diagrams containing them represent specific geometrical isomers or not may be stated in the text of the paper. If nothing is said, the nature of the chemistry may indicate the answer to a chemist. For example, it might be reasonable in the absence of information to the contrary to assume that a diene created by oxidative ring opening of a benzene-1,2-diol or benzene-1,2-diamine by lead tetraoxide has *cis,cis* configuration at the double bonds with respect to the carbon atoms that made up the ring. Even that might be open to debate, though, given the capacity of double bonds to rotate if there are opportunities for tautomerism or mesomerism, so what is the abstractor to do if the authors say nothing one way or the other? Building databases is time consuming and frequently done against tight deadlines. How much time should the abstractor spend finding out the facts, if they can be found? Arbitrarily drawing crossed bonds could lead to the loss of valuable information about a particular isomer or how to make it. Taking the drawn structures at face value is the safer option and the one that is usually preferred.

Crossed bonds were handled badly by low resolution computer graphics. It was not easy to tell whether you were looking at a badly drawn crossed bond or a badly drawn normal double bond. The supplier of at least one chemical reaction database decided simply to draw all structures with the largest substituents at double bonds trans to each other unless there was specific information to the contrary. Given the uncertainties over what is published anyway, this is not such a shocking decision, but users unaware of it are likely to find puzzling surprises.

A different consideration of style provides fun for developers of software using the crossed bond representation if they are interested in things like reaction prediction. If a reaction can lead to the formation of both geometrical isomers a crossed bond will be put into the drawing of the product by the system. When the double bond is in a chain there is no problem, but if the double bond is in, say, a five or six-membered ring it is obvious to a chemist that the atoms in the ring must be *cis* to each other. It will be irritating to see the system determinedly drawing crossed bonds. Strictly speaking, it will be wrong as well as irritating: the geometry must be specific when the double bond is in a fairly small ring, and it is not correct to suggest otherwise. So a conscientious software developer needs to build checks into the code to decide when to generate crossed bonds and when to be specific even for the same reaction.

8.2.3 Other Kinds of Asymmetry

Some computer applications can handle square planar, pentagonal, and hexagonal stereochemistry but not many do. The big market for software has been in day-to-day organic chemistry where these kinds of asymmetry are not often encountered. Gradually, the gaps in what software can handle are being filled but it is likely to be a long time before rotamers and similarly unusual kinds of isomers are covered by everyday chemical information software, if they ever are.

References

1. W. T. Wipke and T. M. Dyott, Simulation and Evaluation of Chemical Synthesis. Computer Representation and Manipulation of Stereochemistry, *J. Am. Chem. Soc.* 1974, **96**, 4825–4834.
2. http://www.mdl.com/products/pdfs/ Enhanced_Stereochemical_Representation.pdf.
3. J. A. Elvidge, P. N. Judson, A. Percival and R. Shah, Preparation of Some Highly Polarised Ethenes by the Addition of Amines to Suitable Carbonitriles, *J. Chem. Soc, Perkin Trans. 1.* 1983, 1741–1744.

DEREK – Predicting Toxicity

9.1 How DEREK Came About

Chemists at Chesterford Park collaborated in the LHASA project for some time before there was a computer at the research station. A colleague, Graham Rowson, and I travelled frequently to Leeds to spend weekends there, writing transforms about heterocyclic chemistry for the knowledge base. We went caving by day in the Yorkshire Dales and worked in the evenings and into the night, as the computer was in heavy demand during normal working hours. But after a year or so, a Digital Equipment Corporation VAX computer was installed at Chesterford Park and we licensed LHASA on site, complete with source code.

The driving force for the installation of the computer was the need for database and laboratory information management systems in the biological screening and toxicology departments. The computer services department, which had hitherto looked after the needs of the finance and business management departments at head office, engaged staff and contractors with knowledge of computer methods in biological information management. For LHASA, staff would need a strong background in chemistry and would have to learn to work in FORTRAN and Pascal – languages which the computer services department did not want to take on. In addition, the use of LHASA would be highly experimental, whereas all the other computer applications would be fully maintained and supported services. It was agreed that a newly-formed chemical information and computing section would have sole responsibility for LHASA and chemistry-related software, and thus it was that we began to develop an understanding of the inner workings of LHASA within the chemistry department.

Derek Sanderson, the head of toxicology, was looking for a solution to a problem. Every week he scanned the structures of chemicals recently submitted

RSC Theoretical and Computational Chemistry Series No. 1
Knowledge-based Expert Systems in Chemistry: Not Counting on Computers
By Philip Judson

Published by the Royal Society of Chemistry, www.rsc.org

for biological screening and made judgments about their potential toxicological hazards. If he did not like the look of a compound he would warn the chemist, recommend early toxicological testing, and discuss ideas for safer analogues. (You might wonder why he did not advise chemists on potential toxicological hazards before they synthesised chemicals, but by not making and testing a chemical you might miss an important new lead. It was only after a lead had been discovered that it became important to take toxicity into account in designing follow-up structures. So it did not much matter whether Derek looked at structures before, or after, they were first sent for testing, and there were practical reasons why it was more convenient to do it after). He would be retiring in a few years and his knowledge would leave the company with him. In any case, would it not be better if the chemists had easy access to advice without having to telephone the head of toxicology?

The computer services department was already in a collaboration to develop a prototype expert system to help with formulating pesticides, and Derek saw that a system could be developed to alert people to toxicological hazards. The formulation project was near to completion, and so an outline proposal was drawn up for a follow-on project to develop an expert toxicity prediction system. It was clear that the formulation system would need further development if it was going to be acceptable to scientists for everyday use. One weakness was that it was, so-to-speak, written from the inside outwards: it communicated with knowledge base developers, and reported the reasons behind its proposals to end-users, in an unappealing LISP-like style (the program itself was written in LISP). Another weakness was that it was very slow. But its biggest weakness was in the way it elicited information from the user. The system needed to know things about the structure of the chemical the user wanted to formulate, but it had no chemical perception module. Instead it asked the user a string of questions. Is the compound aromatic? Has it got any chlorine atoms in it? Is it a carboxylic acid? I sent a memo to Derek and the team preparing the proposal for a toxicology expert system, suggesting that we could base the toxicity prediction system on LHASA.

A week or two later, David Evans, the director of research, came into my office clutching a memo from the main board in Berlin. They had seen the formulations prototype and were unimpressed. The memo said that the project had been terminated and that no resources were to be put into new expert system projects. During lunch hours over the ensuing couple of weeks, I created a primitive system based on LHASA for recognising toxicophores in structures, highlighting them, and issuing simple textual statements such as "Potential cholinesterase inhibitor" when presented with the structure of an appropriate organophosphorus compound. When members of the main board were next at Chesterford Park, David brought them to see our demonstrator. They were favourably surprised. They were more than that. To our alarm in the Chemical Information and Computing Section, they were positively enthusiastic and looked to us to develop a more advanced prototype.

Paul Hoyle, then completing his PhD with Peter Johnson in Leeds, came and helped with some of the program modifications. Chris Earnshaw, who worked with me in the Chemical Information and Computing Section, liaised with Derek Sanderson to put together a first, small knowledge base, as well as working on a lot of the programming. The only remaining problem was what to call the new system. Within the chemical information and computing section we had jocularly dubbed it "The Electric Derek", since it was supposed to emulate Derek Sanderson. We toyed with making that its official name, but we were not quite brave enough. We settled on simply calling it DEREK, but we felt we had to have a suitably staid reason even for that name, and so "D E R E K" was written large on a sheet of paper on the office wall and everyone was invited to suggest acronyms. We ended up with "Deductive Estimation of Risk from Existing Knowledge". We should have been ashamed to go forward with so contrived a justification for the name, let alone to have used the word "risk" for a system designed to recognise hazard, not risk, and later we became so. "Deductive Estimation of Risk from Existing Knowledge" has, we hope, been consigned to the cutting room floor of fate, and Derek for Windows, which superseded DEREK, is overtly named after Derek Sanderson, whose thoughts on mechanisms of toxicity sit still at the core of its knowledge base.

Popular advice has it that the worst possible reason for building an expert system is to try to capture the knowledge of someone about to retire. Perhaps, in the light of our experience, that reason is not such a bad one, as long as it is not where things stop. Derek Sanderson had been keen from the start to get access to knowledge from other toxicologists and add it the system, and that was what happened. Soon after we had completed the prototype, Derek Sanderson and I gave talks about it at a meeting of the British Toxicological Society.[1,2] The management at Schering AG agreed that knowledge sharing was the way forward. Our licence for LHASA allowed us to modify and to use the source code internally, and only internally, but encouraged us to make the results of work available to the team at Harvard and their collaborators. Schering donated DEREK back to the LHASA team at Harvard and to Lhasa Limited, on condition that Lhasa Limited set up a knowledge sharing scheme to develop DEREK, similar to the one operating for LHASA knowledge base development. At about that time, the first peer-reviewed paper about DEREK was published.[3]

Initially, collaborative work focussed on covering the so-called "Ashby and Tennant alerts" – Ashby and Tennant had published a paper[4] in which they drew a now-famous, hypothetical molecule containing all the substructural features, "alerts", that were believed to make molecules potentially mutagenic. Some work was done, though, on other end points, using knowledge that collaborators offered to donate. As work progressed and companies became more comfortable about knowledge sharing, many made substantial donations to the project, both in cash and in kind. In most cases, their contributions are acknowledged in comments in the knowledge bases of DEREK and Derek for Windows (about which, more later) and in various papers and posters, but I will not list them here. Some donations have been made by organisations who prefer not to be publicly named. It would be unfair to name some but not all of the donors. Let it suffice to say that donations in cash and kind over the years, by sponsoring organisations, including important proprietary knowledge, have gone to make up the greater part of the knowledge base. A paper describing further collaborative progress was published in 1996,[5] and there have been many other papers and posters about the program and its successor, Derek for Windows.[6]

9.2　The Alert-based Approach to Toxicity Prediction in DEREK

The basic assumption behind the prediction of toxicity by a human expert such as Derek Sanderson or a computer expert system such as DEREK, is that the biological activity of a compound is determined by its structure. It seems a reasonable assumption, and it is supported in practice. Toxicity may be due to specific interaction with a biological molecule causing a change in its behaviour, or through non-specific disruption of the structure of a cell or the chemistry operating in it. In either case, someone having the right knowledge and given the structure of a compound can make predictions about its potential toxicity.

Specific interactions are often called "lock and key" mechanisms, where a toxic molecule has the right shape to fit a protein site and has features such as hydrogen-bonding centres in the right places for tight binding to the site. Becoming bound to the site of action of an enzyme, it prevents the normal substrate from entering. Becoming bound to a protein site intended to respond to a hormone, it may falsely trigger the response that the hormone triggers or disable the protein and prevent the hormone from acting, depending on how closely its structure resembles that of the hormone – that is, it may behave as an agonist or antagonist. Or the toxic molecule may become bound to an enzyme in a place remote from its active site, causing the protein to change shape, or interfering with its flexibility, in a way that prevents the active site from functioning – so-called allosteric interaction.

A much-quoted example of lock-and-key interaction, and one that was covered in the first rules to be written for DEREK, is the toxicity of the organophosphorus acetylcholinesterase inhibitors. These are the notorious nerve gases and related insectides. Acetylcholinesterase, as its name suggests, catalyses the saponification (hydrolysis) of acetylcholine, Structure 9.1 (see Figure 9.1). Acetylcholine acts as a messenger between one nerve and the next, and so disrupting its removal after the signal has been passed on spells disaster for nerve communications.

Acetylcholinesterase catalyses the hydrolysis by stabilising the reaction intermediate and thus lowering the energy barrier for the reaction, as illustrated in Figure 9.1. A phosphate group is also tetrahedral and has hydrogen bond acceptors spatially arranged in the same way as the ones in the intermediate in the ester hydrolysis, Structure 9.2, but unlike the intermediate in the hydrolysis

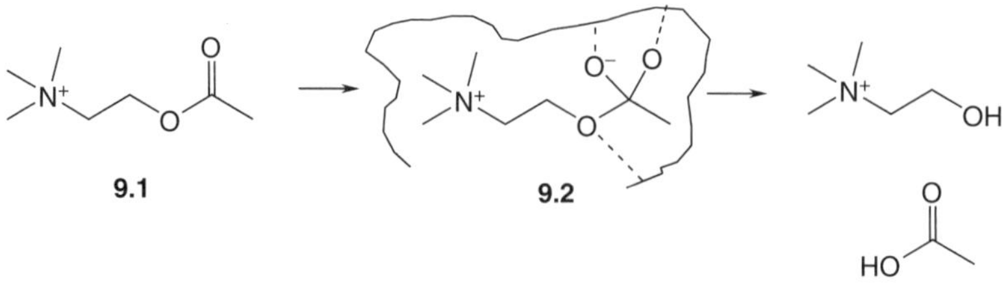

Figure 9.1 Stabilisation of the reaction intermediate by acetylcholinesterase.

of an ester, it is stable and so once it is bound to the enzyme site there is no mechanism for its spontaneous removal. If the molecule containing the phosphate group is like acetylcholine in other ways as well, it binds strongly. The site is blocked, and the enzyme ceases to function.

Omethoate (see Structure 9.4) – formed when dimethoate (see Structure 9.3) is metabolised – is such a compound. The N-methylamide group in dimethoate and omethoate may not appear a good mimic for the trimethylated quaternary amine in acetylcholine. However, amides are strongly polarised and the nitrogen atom has a significant positive charge, creating a positively-charged centre with an admittedly small lipophilic substituent, the methyl group, to mimic the trimethylammonium group of acetylcholine.

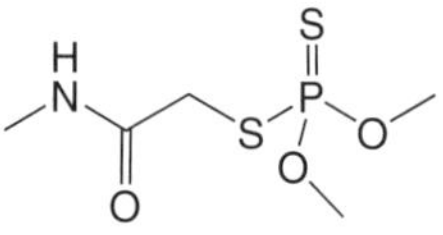

9.3 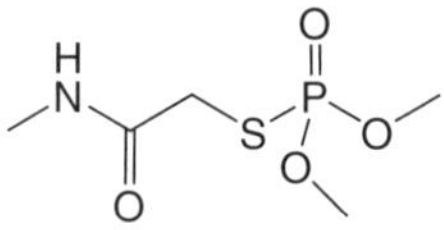   **9.4**

A human expert, or a knowledge-based computer system, can predict the potential for a compound to inhibit acetylcholinesterase, on the basis of some rules. If a structure contains a phosphate group, it is likely to be an acetylcholinesterase inhibitor. If in addition, it contains a nitrogen atom, or some other functionality, that will be positively charged at biological pH, about 5.2 Å away from the centre of the phosphate group (or, expressed differently, four bonds away) and with small lipophilic groups attached to it, the compound is likely to be strongly active. More generally, structures likely to be toxic might be represented by Structure 9.5, where R^1, R^2, R^3, and R^4 are simple hydrocarbon substituents, Y can be C, O, or S, and X can be O or S. An accompanying rule might state that when X is S, metabolic activation is believed to be necessary.

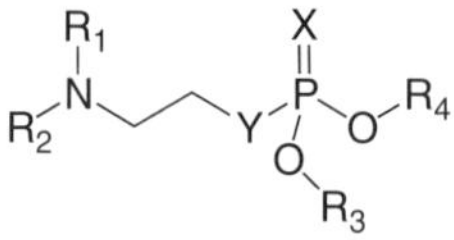

9.5

Living cells are highly susceptible to damage through lowering of the pH of their contents or their surroundings, and so acids are examples of compounds with a non-specific mode of toxic action. No direct interaction with a particular biological molecule is involved. The change in pH causes general mayhem, probably by upsetting the physicochemical balance at interfaces between aqueous and non-aqueous phases, causing biological membranes to collapse, as well as disturbing a host of pH-sensitive reactions essential to the normal functioning of cells. So it is possible to write a rule that compounds containing an acidic group such as a sulfonic acid group, $-SO_3H$, are likely to be corrosive or irritant, the intensity of the activity depending upon the environment of the group in the structure and the physical properties of the compound. A well-designed rule will take account of the fact that it is the effect of the compound on aqueous pH that is important, and that therefore the harmful potential of an acidic group in a compound may not be realised if the compound also contains a basic group. One could, of course, write a rule based on pK_a instead, but looking for acidic groups and other features in a structure is sufficient for practical purposes and avoids the need to measure or calculate pK_a.

The first rule to be written for the prototype that became DEREK, is another example of a non-specific mode of action. It predicts the activity of α-haloketones such as bromoacetone (1-bromopentan-2-one) as lachrymators, or eye irritants. The reason for their activity is that in contact with tear fluid, they release corrosive acids which intensely irritate the eye. Bromoacetone, for example, releases hydrobromic acid.

Modifying LHASA to use it for toxicity prediction was a combination of disabling the greater part of the LHASA code and making cosmetic changes. None of the code for applying synthetic strategies or generating precursors for transforms was needed. Words in the user display needed to be changed to refer to toxicological endpoints, and the graphics needed to be changed to display the query structure with the substructural feature of interest in it highlighted. The chemical perception modules were retained, and the code for recognising retrons in LHASA recognised substructures associated with toxicity in DEREK. The rules were written in PATRAN and CHMTRN. The sub-structure to trigger the α-haloketone rule for lachrymation, for example, could be written in PATRAN as:

Br,Cl,I-C[HETS=1]-C(=O)-C

Substructures believed to be responsible for toxicological activity are termed "toxophores" or "toxicophores", from the Ancient Greek "toxicon", "arrow poison" and "phor" "to carry". In early versions of DEREK we used the word "toxophore" but my preference now is for "toxicophore", which is widely used in English language publications. "Toxophore" has a long history of use but according to my dictionary "toxon" was the Ancient Greek for arrow and "toxicon" for arrow poison, and so "toxicophore" seems more satisfactory.

Actually, there is room for argument about using either term for some of the substructural features that toxicologists associate with toxicity, depending on what you intend the term to mean. It seems safe to call Structure 9.5 in which X is oxygen a toxicophore, but what about cases where X is sulfur? Such compounds may have some activity in themselves, but it is probable that they are converted metabolically to the oxygen analogues and that it is those that bind tightly to the enzyme site, leading to high activity.

It may be pointless and pedantic to argue about whether the sulfur analogues should be called "toxicophores" or whether the term should really attach only to their oxygen analogues. But what about *n*-hexane? *n*-Hexane is remarkable for being, alone among the straight chain hydrocarbons, capable of causing nerve damage in humans. When we were seeking sponsorship early in the development of DEREK, a potential sponsor challenged us show that a computer system could predict this activity. It is hard to see how a computer algorithm based on some kind of juggling with numbers could do so, but once you are aware of a theory about the mechanism of the toxicity, it can easily be described in a knowledge-based system. A metabolic process oxidises methylene groups adjacent to terminal methyl groups in hydrocarbon chains to hydroxyl groups. So hexane can be converted to hexane-2,5-diol. Further metabolic oxidation of alcohols leads to ketones, generating hexan-2,5-dione in this case. A 1,4-dione, which is what hexan-2,5-dione is, can react with primary amine groups, in protein side-chains for example, to form stable pyrroles through dehydration (see Figure 9.2). The theory is that this is the event with toxic consequences. Now the uniqueness of hexane is explicable. The same sequence of reactions starting even from *n*-heptane or *n*-pentane, seemingly so similar to *n*-hexane, would not lead to the formation of stable, aromatic rings. Assuming the proposed mechanistic explanation to be correct, the rule can be extended to cover more than just *n*-hexane – 3-hydroxyheptane, for example. Indeed there are other compounds to support the theory. The rule in the current version of Derek for Windows is based on both Structures 9.6 and 9.7, with constraints on the R-groups such as that R^2 and R^3 may not be electron withdrawing groups (which may inhibit dehydration to form a pyrrole) and must not be too bulky.

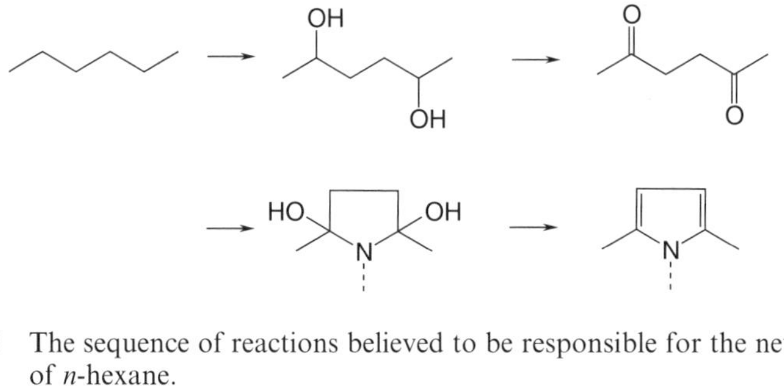

Figure 9.2   The sequence of reactions believed to be responsible for the neurotoxicity of *n*-hexane.

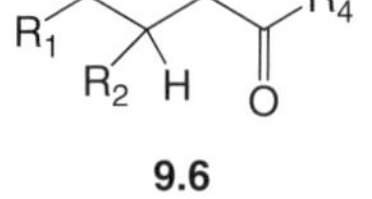

9.6

$$R_5 \quad R_3 \ H$$

9.7

The name most favoured among toxicologists for a substructural feature that should cause toxicological concern is an "alert" – the term used, for example, by Ashby and Tennant in their paper about features associated with mutagenicity.[4] It has the advantages of being more immediately understandable than "toxicophore", more expressive of its primary purpose of alerting someone to a potential problem, and more appropriate for structures like *n*-hexane that give cause for concern, but are not the structures actually responsible for a toxicological effect. Probably a third of the alerts currently contained in Derek for Windows are features that may be converted into true toxicophores by metabolism, rather than that confer toxicity directly. The difference can be observed in assays like the Ames test for mutagenicity,[7] where a substance may show activity only in the presence of a liver extract to bring about metabolic conversions, but for many alerts and end points there is no direct experimental evidence one way or the other. All that is known is that a high proportion of compounds containing the alert are active. The mechanism of action is not necessarily a mystery, but it may be a matter of theory rather than empiricism.

The alert for *n*-hexane and related compounds illustrates a fundamental difference between knowledge based systems and systems based on automated learning or mathematical models: a knowledge-based system is simply a place in which to store human knowledge. That does not mean it cannot make predictions. Like a human expert, it applies existing knowledge to new problems. For its predictions to be successful the knowledge must be general but soundly based, and so a pre-requisite for writing good alerts is that there must be an understanding of the mechanism of toxicity. Ideally, such an understanding would be based on experimental evidence but that is rarely available. More usually, it depends on a satisfactory rationalisation based on theory about the likely mechanism. That being the case, like a human expert, the computer system must be able to explain the reasons for its predictions so that a user can make his or her own judgments about them.

References

1. D. M. Sanderson, *Computer Prediction of Possible Toxicity from Chemical Structure*, presented at the autumn meeting of the British Toxicological Society, University of Newcastle-upon-Tyne, 20–22 September 1989.
2. P. N. Judson, *The Use of Expert Systems for Detecting Potential Toxicity*, presented at the autumn meeting of the British Toxicological Society, University of Newcastle-upon-Tyne, 20–22 September 1989.

3. D. M. Sanderson, C. G. Earnshaw and P. N. Judson, Computer Prediction of Possible Toxic Action from Chemical Structure; the DEREK System, *Hum. Exp. Toxicol.*, 1991, **10**, 261–273. [Note: P. N. Judson was omitted from the author list in the original paper. An erratum was issued subsequently by the publisher, but library catalogues *etc.* usually list the paper in the names of the first two authors only].

4. J. Ashby and R. W. Tennant, Chemical structure, *Salmonella* Mutagenicity and Extent of Carcinogenicity as Indicators of Genotoxic Carcinogenisis Among 222 Chemicals Tested in Rodents by the US NCI/NTP, *Mutagenesis*, 1988, **204**, 17–115.

5. J. E. Ridings, M. D. Barratt, R. Cary, C. G. Earnshaw, E. Eggington, M. K. Ellis, P. N. Judson, J. J. Langowski, C. A. Marchant, M. P. Payne, W. P. Watson and T. D. Yih, Computer Prediction of Possible Toxic Action from Chemical Structure: an Update on the DEREK System, *Toxicology*, 1996, **106**, 267–279.

6. There is a list of publications about DEREK and Derek for Windows on the web site of Lhasa Limited, at www.lhasalimited.org, currently to be found *via* the "Research" page on that site.

7. B. N. Ames, F. D. Lee and W. E. Durston, An Improved Bacterial Test System for the Detection and Classification of Mutagens and Carcinogens, *Proc. Natl. Acad. Sci.*, 1973, **70**, 782–786.

Other Alert-Based Toxicity Prediction Systems

10.1 TOX-MATCH and PHARM-MATCH

DEREK was not the first knowledge-based system to be developed that predicted toxicological hazard from the presence of alerts, or toxicophores, in structures. TOX-MATCH, developed by Joyce Kaufman, Walter Koski and colleagues pre-dated it by several years.[1]

TOX-MATCH was almost identical in concept to DEREK. The user drew the compound of interest. TOX-MATCH looked for matches with structural fragments described in its knowledge base and if it found any it alerted the user to the potential toxicological hazard. The knowledge base language was written using character strings similar to PATRAN. I carried out a survey of computer methods for predicting toxicity for the United Kingdom Ministry of Agriculture, Fisheries and Food in 1991 and went to see Joyce Kaufman in Baltimore. By then, she and Walter Koski had both retired but, in the manner of many retired researchers, they still went to work. TOX-MATCH and PHARM-MATCH, however, were no more. They had been developed to run on hardware which had become obsolete, and written in a language specific to it. The programs were on the one machine of its kind that still remained at the university, but that machine was out of order and unlikely to be repaired.

Joyce Kaufman told me how she had given many presentations about PHARM-MATCH and TOX-MATCH, but no-one had shown any interest in using them or sponsoring the research. So the project had been abandoned some time before my visit. Koski and Kaufman came up with PHARM-MATCH and TOX-MATCH too soon and too late: they developed their prototypes at a time when the opportunities for demonstrating software were limited, and before the toxicology community was ready to take the idea of

RSC Theoretical and Computational Chemistry Series No. 1
Knowledge-based Expert Systems in Chemistry: Not Counting on Computers
By Philip Judson
© Philip Judson 2009
Published by the Royal Society of Chemistry, www.rsc.org

computer prediction seriously; by the time the user community was interested, it was too late in the careers of the inventors – they had already retired.

It is curious that the use of knowledge-based systems to store information about pharmacophores (substructural features associated with pharmacological activity) – the purpose of PHARM-MATCH – does not seem to have been taken up. When I worked in agrochemical research, a book was kept in the chemistry department office, in which were recorded the definitions of "active series" – hand-drawn Markush diagrams representing groups of compounds that had been, or were being, synthesised because of their promising agrochemical activity. Including Structure 10.1 here by way of an example, gives nothing away that should not be, since it is the Markush structure published in a patent claim for fungicides.[3] Chemists had to depend on the memories of long-serving staff or to leaf through the book to make sure they did not re-invent things already discovered and on record.

10.1

R^1= H, alkyl, alkenyl
R^2= alkyl, aryl
R^3= H, alkyl, alkenyl
R^4= CO_2R^5, $CONR^6R^7$
R^5, R^6, R^7= H, alkyl

I left Chesterford Park soon after the development of the prototype DEREK and so I do not know whether it was used to store the information from the active series book in a more accessible form. However, I was involved on behalf of Lhasa Limited in discussions with another company about using DEREK for that purpose. The project did not go ahead, but it was not because the software would not have been able to store and communicate the required information. Everything that the company needed to record could have been recorded, and all of the generic structures that the company used to define active series could be described in PATRAN and CHMTRN. The only program changes needed were minor cosmetic ones – to talk about "pharmacophores" instead of "toxicophores", for example. The deciding factor was the nature of PATRAN and CHMTRN. Staff would need to be trained to write in the languages, and entering the historical information about active series would have used more staff resources than the company considered to be cost effective.

Modern programs like Derek for Windows, which is described in Chapter 13, have simple, graphical knowledge base editors and setting up a knowledge base of pharmacophores would be easy. But in the meantime, many companies have made do with their chemical structure database systems to store information about pharmacophores. That is not a good way to do it, since what is needed is a system for superstructure searching (see Chapter 6.5) – matching stored substructures against full structure queries – whereas chemical structure database systems are designed for substructure searching – matching a query substructure against full structures in the database. You cannot draw a complete molecule and have pharmacophores in the database that are present in your query automatically identified by a system designed for substructure searching. Instead, you have to enter the substructural representation of the pharmacophore you think you might have. In effect, you have to guess what pharmocophores might be in the database.

Over the same period, systems for working with Markush representations have been developed to support the patent information industry, and they may also provide a way of storing information about pharmacophores. Even so, it is surprising that with Derek for Windows in use in over a hundred organisations, no-one seems to have made this obvious alternative use of it (unless they have done so but have seen no reason to make the fact public).

10.2   Oncologic

Oncologic, developed by staff at the US Environmental Protection Agency,[4-6] uses the concept of toxicophores but it is not strictly a knowledge-based system as defined in this book. It is driven by decision trees. A question and a set of valid answers are associated with each node in a decision tree, one answer for each branch at that node. The computer follows a path through the tree directed by the answers. In Figure 10.1 you will find a decision tree designed to mimic part of the task of a waiter. Note the limitations imposed by the tree. The designer has assumed that no civilised person would ask for milk in china tea. If anyone does, the waiter will simply look blankly at them and repeat the only available question at the node following a decision to take china tea – "Lemon?" That is not to say that a decision-tree based system is inadequate – only to illustrate that it is relatively inflexible. The plus side, of course, is that the system developer controls exactly what it will do in every circumstance.

An interesting feature of Oncologic is that it includes a module for making predictions about the potential carcinogenicity of fibres, based upon particle size, shape, and surface properties, rather than chemical constitution. Most other toxicity prediction systems are driven primarily by chemical structure, although they may modify their predictions on the basis of physical properties. There is a module for making predictions about polymers – another area not covered by other systems – one for metals, and one for organic compounds.

The limitations imposed by a decision tree system are apparent in different ways in the metals module and the organics module. By way of illustration, if you select the metals module you are asked whether your metal is radioactive. If you answer "yes" you are told that the system is not designed to give advice about radioactive materials, which is fair enough. However, if you answer "no" you are presented with a list of metals from which to make your selection and, curiously, the designers have included plutonium and some other radioactive metals in the list. If you select plutonium, the program happily proceeds with further questions. The program cannot make assumptions about whether most elements are radioactive or not – if the user is interested in strontium, for example, that might be as ^{88}Sr or ^{90}Sr – so the user is asked whether the query

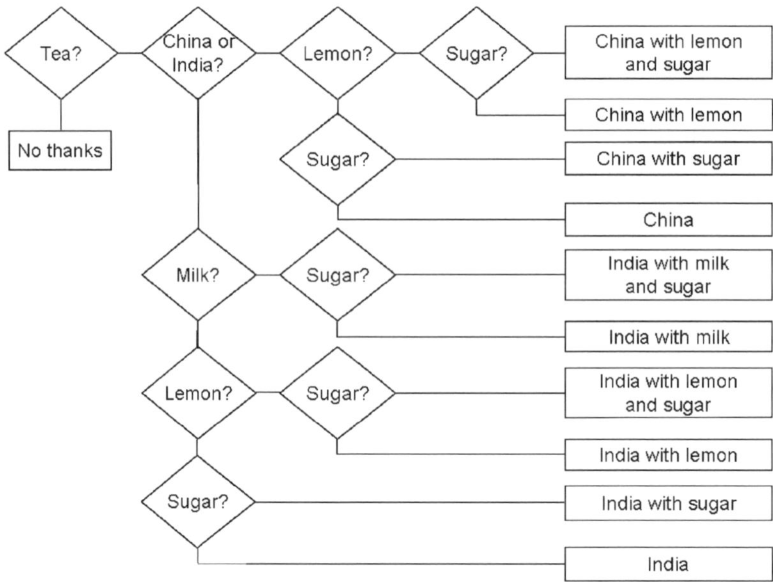

Figure 10.1   A decision tree for a waiter serving tea.

material is radioactive. Including plutonium in the list of non-radioactive metals you can choose from might be a mistake or it might be deliberate on the grounds that you cannot be sure no-one will come up with a non-radiactive isotope of plutonium one day.

You are only allowed to build structures for the organics module from a pre-determined set of fragments – the ones that are associated with nodes in trees. In consequence you are rarely able to draw, and hence to get advice on, the structure you are actually interested in. The absence of a chemical perception module and automatic mapping of alerts from a knowledge base to the query structure, also mean that Oncologic shares the weakness of the system for advising on potential formulations for pesticides mentioned in Chapter 9 – the user is taken through an irritating series of questions.

Oncologic was developed very early in the history of toxicity prediction. It pushed the limits of the technology of the time, and its designers had the imagination to see that a computer could predict toxicity by a process of rules and logic when the majority of researchers were focussed on trying to develop mathematical models. Oncologic remained unchanged for a long time and its origins as a DOS application are apparent in the style of its user interface, but it is available for download free of charge from the EPA website.[7] Rumour has it that work is in progress on developing a new version and it may appear soon.

10.3   HazardExpert

HazardExpert, from Compudrug,[8,9] is similar in the way that it works to TOX-MATCH and DEREK. A knowledge base contains descriptions of toxicophores in a language that has similarities to the PATRAN and CHMTRN used in DEREK. It has a particular feature that is unique and that you may judge to be a strength or a weakness.

HazardExpert gives you a numerical estimate of the probability that your compound will be toxic against the endpoint or endpoints you have chosen. Expressed a bit simplistically this is how it works. People compiling information for alerts in the knowledge base have access to data for a large number of compounds. Usually, not all compounds containing an alert are active. Steric hindrance, the presence of other structural features that counter the action of the alerting substructure, and physicochemical properties all have their influence. Compounds containing some alerts are pretty well all toxic. At the other extreme, only very few compounds containing some other alerts are toxic. So to give guidance, a probability of activity is associated with each alert in the knowledge base, derived from the proportion of compounds in the training set that were active. If one in ten compounds containing the alert in the training set is active, then the probability that a novel compound containing the alert will be active is 0.1.

So what about a query compound that contains two different alerts? If I understand correctly, the probability that it will be active is computed in accordance with standard probabilistic arithmetic. So if the probability of toxicity on account of the presence of the first alert is 0.5 and the probability on account of the second alert is 0.4, then the probability, P_A, that the compound will be toxic is

$$P_A = 0.5 + 0.4 - 0.5 \times 0.4$$
$$= 0.7$$

My problem with this, is that the mathematics of probability are based on the laws of chance. The biological activity of a compound is not a chance event. There are mechanistic reasons why a structure containing a particular substructure is, or is not, active. If a structure contains more than one toxicophore there is no apparent mechanistic reason for supposing that the overall likelihood of activity can be determined by applying the laws of chance. I am uncomfortable about the probabilistic approach in HazardExpert, both for attaching a probability to each alert and for combining probabilities when more than one alert is present, and will touch on the subject again in Chapter 13, but you may disagree.

10.4 BfR/BgVV System

A team working initially at the Bundesinstitut für Gesundheitlichen Verbraucherschutz und Veterinärmedizin (BgVV) in Germany and later at the Bundesinstitut für Risikobewerkung (BfR), the German Federal Institute for Risk Assessment, developed a system for predicting skin and eye irritancy and corrosivity from chemical structure and published the rules.[10–12] Some of the rules are based on structural alerts, but whether they apply for a particular query structure takes into account broader information about the structure, such as whether it is composed only of carbon, hydrogen and oxygen, or contains other elements. The rules are easily expressed in the form of decision trees but the application that the team developed perceived features in structures for itself – the user did not have to answer a string of questions triggered by the decision trees. Entering structures for processing required typing in a linear code that was peculiar to the application and this did not appeal to users. The rules, however, were, and are, of interest and, being published, they have been incorporated into other applications. For example, they are available in ToxTree (see Section 10.5).

10.5 ToxTree and Toxmatch

ToxTree was developed for the European Chemicals Bureau by IdeaConsult[13] and it can be downloaded free of charge from the European Chemicals Bureau.[14] It is a simple decision tree system and is not primarily intended for toxicity prediction. It incorporates several classification schemes to help with deciding, for example, which QSAR (Quantitative Structure–Activity Relationship) models are most suitable for a given compound, and that is more its intended purpose. However, there are some toxicity prediction rules in it, including the rules about skin and eye irritancy and corrosivity developed at the BgVV and BfR (see Section 10.4).

Not to be confused with TOX-MATCH (see Chapter 10.1), Toxmatch[15], which was also developed by IdeaConsult[13] and can be downloaded free of charge from the European Chemicals Bureau,[16] helps with the categorisation of chemicals, like ToxTree, and is not a toxicity prediction system. It provides methods for calculating a variety of physico-chemical descriptors such as log P, ionisation potential, and molecular surface area and for grouping chemicals according to their similarity.

10.6 Environmental Toxicity Prediction

Various researchers have described work on alert-based prediction of human or environmental toxicity, which have not, to date, led to the commercialisation of widely available computer applications. Gerrit Schüürmann's group in Leipzig, for example, have reported work on alerts for predicting excess toxicity (*i.e.* greater than would be expected from narcosis) to daphnia.[17] There is more about this in Chapters 11.4.2 and 17.7.

References

1. J. J. Kaufman, W. S. Koski, P. Harihan, J. Crawford, D. M. Garmer and L. Chan-Lizardo, Prediction of Toxicology and Pharmacology Based on Model Toxicophores and Pharmacophores using the New TOX-MATCH-

PHARM-MATCH Program, *Int. J. Quantum Chem., Quantum Biol. Symp.*, 1983, **10**, 375–416.

2. W. S. Koski and J. J. Kaufman, TOX-MATCH/PHARM-MATCH Prediction of Toxicological and Pharmacological Features by Using Optimal Substructure Coding and Retrieval Systems, *Anal. Chim. Acta*, 1988, **210**, 203–7.

3. G. P. Rowson, A. Percival and P. N. Judson, *Fungicidal Cyanopropenoates and Compositions Containing Them*. Eur. Pat. Appl. EP 88 545, 14th September 1983, Brit. Appl. 82/6480, 5th March 1982.

4. Y.-T. Woo, D. Y. Lai, M. F. Argus and J. C. Arcos, Development of Structure Activity Relationship Rules for Predicting Carcinogenic Potential of Chemicals, *Toxicol. Lett.*, 1995, **79**, 219–28.

5. D. Y. Lai, Y.-T. Woo, M. F. Argus, and J. C. Arcos, Cancer Risk Reduction Through Mechanism-Based Molecular Design of Chemicals, in *Designing Safer Chemicals*, ed. S. De Vito and R. Garrett, ACS Symposium Series Vol. 640, Am. Chem. Soc., Washington DC, 1996, pp. 62–73.

6. Y.-T. Woo and D. Y. Lai, OncoLogic: a Mechanism-Based Expert System for Predicting the Carcinogenic Potential of Chemicals, in *Predictive Toxicology*, ed. C. Helma, Marcel Dekker, New York, 2005, 385–413.

7. http://www.epa.gov/oppt/newchems/tools/oncologic.htm

8. M. P. Smithing and F. Darvas, HazardExpert – an Expert System for Predicting Chemical Toxicity, in *Food Safety Assessment*, ed. J. W. Finley, S. F. Robinson and D. J. Armstrong, ACS Symposium Series Vol. 484, Am. Chem. Soc., Washington DC, 1992, pp. 191–200.

9. HazardExpert comes from CompuDrug International Inc., 115 Morgan Drive, Sedona, AZ 86351, USA.

10. J. D. Walker, I. Gerner, E. Hulzebos and K. Schlegel, The Skin Irritation Corrosion Rules Estimation Tool (SICRET), *QSAR Comb. Sci.*, 2005, **24**, 378–384.

11. I. Gerner, S. Zinke, G. Graetschel and E. Schlede, Development of a Decision Support System for the Introduction of Alternative Methods into Local Irritancy/Corrosivity Testing Strategies. Creation of Fundamental Rules for a Decision Support System, *Alternat. Lab. Animals*, 2000, **28**, 665–698.

12. I. Gerner, M. Liebsch and H. Spielmann, Assessment of the Eye Irritating Properties of Chemicals by Applying Alternatives to the Draize Rabbit Eye Test: the Use of QSARs and In Vitro Tests for the Classification of Eye Irritation, *Alternat. Lab. Animals*, 2005, **33**, 215–237.

13. Ideaconsult Limited., 4 Angel Kanchev Street, 1000 Sofia, Bulgaria.

14. http://ecb.jrc.ec.europa.eu/

15. G. Patlewicz, N. Jeliazkova, A. G. Saliner and A. P. Worth, Toxmatch - a New Software Tool to Aid in the Development and Evaluation of Chemically Similar Groups, *SAR QSAR in Environ. Res.*, 2008, **19**, 397–412.

16. http://ecb.jrc.it/qsar/qsar-tools/index.php?c=TOXMATCH

17. P. C. Von der Ohe, R. Kühne, R.-U. Ebert, R. Altenburger, M. Liess and G. Schüürmann, Structural Alerts – a New Classification Model to Discriminate Excess Toxicity from Narcotic Effect Levels of Organic Compounds in the Acute Daphnid Assay, *Chem. Res. Toxicol.*, 2005, **18**, 536–555.

CHAPTER 11

Rule Discovery

11.1 QSAR

Expressed in its most generalised way, statistical QSAR modelling assumes that biological activity can be determined by applying a mathematical function to a set of numerical descriptor values. One of the most well known is the Hansch equation, see eqn (1), proposed by Corwin Hansch and co-workers for calculating biological activity on the basis of the attributes of a variable substituent in a chemical structure:[1,2]

$$\log(1/C) = a\pi + b\pi^2 + c\sigma + dE_S + k \tag{1}$$

where

π is a hydrophobic term,

σ is an electronic term,

E_S is a steric term,

and   a, b, c, d, and k are constants.

The measure chosen for the hydrophobic term is most usually the octanol/water partition coefficient, $\log P$. It is convenient to measure octanol/water coefficient experimentally as a surrogate for partition between water and fatty membranes in living cells. Corwin Hansch's group showed that $\log P$ could be estimated by summing contributions from substructural fragments[2] and developed the widely used program, $C\log P$ (ref. 3) and $\log P$ is more often calculated than measured. A variety of ways to calculate values for σ and for E_S have been described, but the "electrotopological states" of Hall *et al.*[4] are widely used for the former, and Taft values[5,6] for the latter.

RSC Theoretical and Computational Chemistry Series No. 1
Knowledge-based Expert Systems in Chemistry: Not Counting on Computers
By Philip Judson

Published by the Royal Society of Chemistry, www.rsc.org

Analyses taking account of contributions to activity from multiple fragments assume that activity is the sum of contributions from the fragments, as in eqn (2). Most commonly the terms are all linear, but sometimes squared terms are also used for some descriptors.

$$\text{Activity} = \beta_0 + \beta_1 X_1 + \beta_2 X_2 + \cdots \beta_n X_n \tag{2}$$

where β_0 to β_n are constants and X_1 to X_n are numerical attributes of the fragments.

Standard mathematical methods are used to solve simultaneous equations for a set of structures with known biological activities and thus to determine the values of the constants, β_0 to β_n. For statistical validity, the variables in the equations need to be independent and this is frequently not the case with properties calculated from substructural fragments. So statistical methods are used to determine which descriptors best correlate independently with activity. The values of β_0 to β_n having been determined, the expected activity for a novel structure can be calculated from the values of X_1 to X_n for the fragments it contains.

There are many books on the subject of QSAR. For convenience, three references given in Chapter 1 are repeated at the end of this chapter.[7-9]

11.2 TopKat

TopKat,[10,11] developed by Kurt Enslein and colleagues, and currently supplied by Accelrys Inc.,[12] makes predictions on the basis of quantitative relationships between substructures from a pre-defined set and toxicological activity. There are several thousand substructures in the library that it uses. Figure 11.1 shows a few examples of the kinds of features you might find in such a library. Numerical attributes are calculated for each substructural feature and these are the descriptors for the analyses. Properties of the whole molecule, such as values encoding information about size and shape, and the estimated octanol/water partition coefficient (log P), are also included as descriptors.

By way of illustration, and not necessarily using fragments that are actually included in the TopKat library, if you entered Structure 11.1 in Figure 11.2 as a query, TopKat might recognise the substructures shown in Figure 11.2b to be

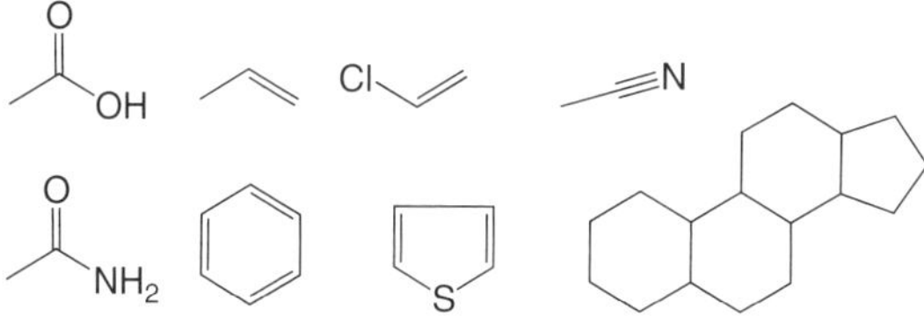

Figure 11.1 Entries that might be found in a fragment library.

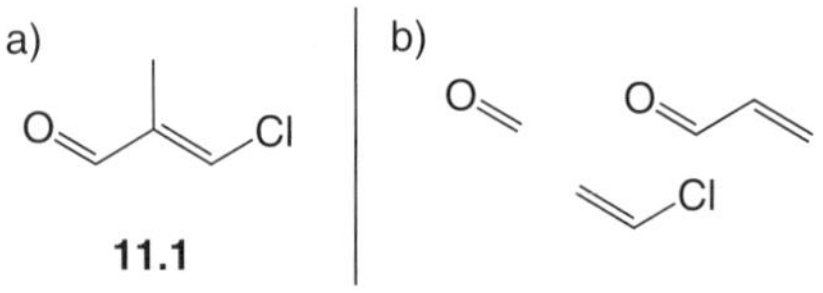

Figure 11.2   Substructural fragments that might be recognised in a query structure.

fragments in its library. If you were looking for information about potential mutagenicity the descriptor values calculated for those fragments would be entered into the mutagenicity equation in TopKat most appropriate to a molecule like Structure 11.1 and the calculated activity would be reported to you. TopKat would also assess, by a procedure that is currently secret but apparently may soon be published in a patent, how well your query fitted into the prediction domain for the chosen equation – *i.e.* how similar your structure is to the ones that were used in the training set to generate the model and taking into account that some regions of chemical space may be covered better than others even within the scope of the training set.

TopKat includes modules for predicting rat acute oral and inhalational toxicity, skin sensitisation, rodent carcinogenicity, mutagenicity, developmental toxicity, and skin and eye irritation. It also covers the environmental endpoints of toxicity to fathead minnow and to daphnia, and it estimates aerobic biodegradability.

Many organic chemists and people in related fields such as toxicology, take too naive an approach to statistics and attach mistaken significance to data or observations, a topic which will come up again in Chapter 18. The rigorous implementation of best practice in statistics in TopKat is rightly emphasised by its developers and suppliers. However, it is increasingly recognised that for a model to be reliable, it must predict for the same toxicological endpoint for compounds that act by the same mechanism. The inclusion of models for, in particular, acute toxicity and developmental toxicity raises some eyebrows, since both comprise a host of different, more specific endpoints involving very different mechanisms of action. It may be that experience shows the models to work, at least in the tests that have been done on them, but to be convinced about any model users need to understand why it works. Be aware also, that no matter how rigorous a statistical analysis is, the validity of the resultant model depends on the appropriateness of the descriptors used in the analysis. Following correct statistical methods is not the whole story.

11.3   Multicase

A snag with the use of a pre-defined set of substructural fragments for modelling, as in TopKat for example, is that it may be biased. Someone decides what to include and what not to include. Recognising this, Gilles Klopman developed a system based on using fragments automatically generated from the

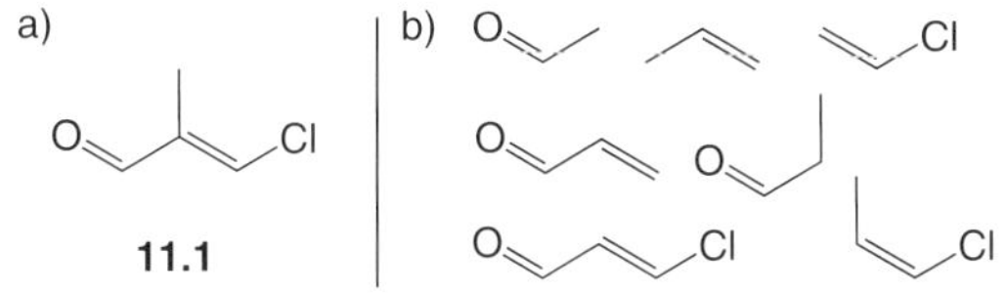

Figure 11.3 Linear fragments in a query structure.

structures in the training set. Subject to constraints on minimum and maximum chain length, they are all the linear fragments it is possible to find in the set of structures.[13] Taking the single Structure 11.1 for illustration, the linear fragments with a chain length of two bonds or more are the ones in Figure 11.3b. The use of these linear fragments underpins the functioning of the Multicase programs, Casetox, M-Case, and MC4PC.[14,15] The method of analysis is statistical. The broad principles have been published but details of how they are implemented remain confidential to the company.

Two further levels of sophistication in the use of linear fragments in Gilles Klopman's programs are worth mentioning. The first is that, to improve discrimination, branching points are flagged in the linear fragments. The second is that two kinds of fragment are recognised: the first kind are fragments that are primarily responsible for the observed biological activity; the second kind are fragments that do not cause activity in themselves, but increase or decrease activity if there is any.

What is found to be associated with activity in a typical analysis, is a small set of linear fragments, not a single one. Gilles Klopman calls this set a "biophore". Once a biophore has been identified, a QSAR can be constructed by statistical analysis of the activities of structures containing the biophore. Partition coefficient, log P, is usually found to be the most significant contributor to variation in activity within a series of compounds with a common biophore.

A large number of prediction modules is available from Multicase Inc. They cover acute mammalian toxicity, hepatotoxicity, renal toxicity, cardiac toxicity, carcinogenicity, developmental toxicity, skin and eye irritation, and more. Some environmental endpoints are covered, including fish toxicity, biodegradability, and bioaccumulation. The module for each of these broad endpoints comprises a set of more specific ones, relating to a single strain and sex of animal, a particular type of symptom or, in some cases, interaction with a particular enzyme. So, while Multicase software covers some endpoints that may seem worryingly vague, they are actually modelled at a more well-defined level.

11.4   Other Fragment-Based Systems

TopKat and the Multicase suite are the fragment-based systems that have become widely known and used. Two others illustrate variations on the same approach. There are more, and I have chosen these two because they are the examples I know something about.

11.4.1 REX

When a biological molecule interacts with a molecule that is toxic it, so to speak, "sees" it from the outside. So there is reason to wonder whether the connecting chain in a linear fragment of the kind used by Multicase is relevant to activity or might even be a distraction. If say the reason why a linear fragment causes activity, is that a carbonyl oxygen at one end of the chain acts as a hydrogen bond acceptor and an amine group at the other end acts as a hydrogen bond donor, it may not matter whether an atom somewhere in the middle of the chain is carbon, oxygen, nitrogen, or anything else. If the training set contains examples in which the atom near the middle differs, the fragment will be missed by an algorithm looking for the consistent occurrence of specific fragments in active molecules.

I did some experiments with a system based on the use of atom pairs, which I called "REX".[16] An atom pair[17] is described by two atom types and the distance between them. Distance might mean through-space distance in three dimensions, but in this context it means the distance expressed as the number of bonds between the atoms. An example might be represented in alphanumeric form as "O3N", meaning an oxygen atom three bonds away from a nitrogen atom. You could take account of bonding by giving atoms a hybridisation attribute (*e.g.* sp, sp^2, or sp^3), but in REX, I chose to allow the ends of descriptor pairs to be atoms or bonds. So, using the # sign to represent a triple bond, #3N would mean a triple bond three bonds away from a nitrogen atom. Some other issues were covered as well, such as what to do about situations where one of several atom types might be acceptable at the end of a descriptor chain (*e.g.* oxygen and nitrogen might both support a requirement for hydrogen bonding, one being found in some structures and the other in others).

The system was non-quantitative. It simply looked for atom pairs that were more common in active compounds than in inactive ones, and suggested that the relationship was significant. The indications were that a computer program could find atom pairs automatically that were associated with activity in training sets of structures, but the research ended without establishing whether they offered any advantages over linear fragments of the kind used in M-Case. Gilles Klopman told me, in an informal conversation, that M-Case can use atom pairs as well as linear fragments and that there seemed to be some evidence that atom pairs worked better for lock and key type activity, while linear fragments worked well for non-specific types of toxicity. It is easy to rationalise this on the grounds that lock and key interactions are of the kind REX was designed for, in which suitably-located centres on an active molecule bind to a biological site, whereas non-specific toxicity is associated with features such as the presence of an acidic group which will be picked up just as well in a linear fragment analysis. Convenient though this rationalisation may be, we have not amassed the evidence to afford it any greater status than a tentative theory. Or, in the interests of exactitude, I do not have the evidence and if Gilles has found it, he has not mentioned it in subsequent conversations.

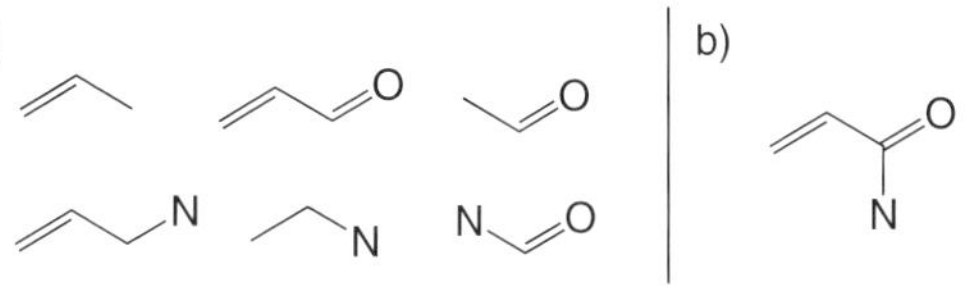

Figure 11.4   A biophore and the toxicophore it might represent.

A perceived weakness with systems like M-Case is that a biophore is a collection of fragments with no specified relationship between them. A real toxicophore is typically a single, branched fragment – comprising the components of the biophore, presumably. Figure 11.4 illustrates this for an acrylamide group – a feature that can be associated with neurotoxicity, chromosome damage, and skin sensitisation. REX allowed the user to view atom pairs mapped onto structures chosen from the training set and to decide whether they should be united to form a more complete toxicophore. One can think of structures that contain the biophore but not the toxicophore, for example Structure 11.2 contains the biophore from Figure 11.4 but not the toxicophore, which suggests that a system based only on biophores would over-predict toxicity. However, the evidence is that over-prediction arising from this cause is not a significant problem; in practice, structures submitted as queries rarely contain complete biophores distributed differently from the way they were distributed in compounds in the training set.

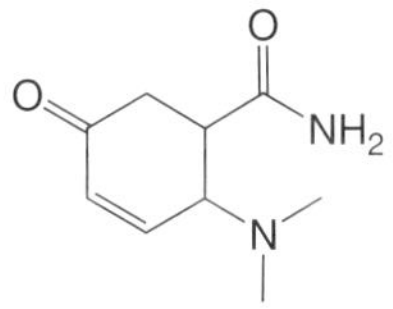

11.2

11.4.2   Using Atom-Centred Fragments

Gerrit Schüürmann's group in Leipzig are among those who have looked at using "atom-centred fragments", also called "augmented atoms", as an alternative to atom pairs or linear fragments.[18,19] They have shown that you can make predictions in the field of ecotoxicity by building statistical models based on atom-centred fragments. One would expect atom-centred fragments also to work for the prediction of mammalian toxicity.

Augmented atom descriptors can be based on all sorts of atom attributes but for the sake of illustration let us consider simply elemental type. Start by labelling all the atoms in a structure according to their elemental type. Now attach a second list of labels to each atom containing the types of its neighbouring atoms. Keep adding lists like this, each time moving out to the set of atoms one bond further away, until you can go no further. More typically the labels include at least information about bond types, as well as atom types, or about the hybridisation states of atoms.

In practice, a limit is placed on how many shells you build around an atom, and it is usually quite a low one – often just two or three. Also, depending on the application and how you design the algorithms that make use of the augmented atom information, you may or may not build and use a set of lists for each atom. In some systems, a single number is generated to represent an atom in a given environment – often a hash code (*i.e.* an algorithmically generated number that is not guaranteed to be unambiguous, but is consistently the same in the same circumstances and can be expected to be different most of the time for different inputs).

11.5   Other Approaches

The LeadScope Predictive Data Miner[20] is a toolbox to support data mining for chemistry-related problems. Graphs and bar charts help you to recognise trends and features in common between structures and the data associated with them. The software includes a large library of structural fragments, different kinds of descriptors and methods for generating them, and tools like the ones used in TopKat and Multicase for building predictive models. In addition, you can enter and store your own structural alerts – some of which the other tools in the package may have helped you to discover.

Outside the scope of this book is a large body of work on applying well-known machine learning techniques to the problem of toxicity prediction. Automated algorithms such as ID3[21] have been used to build decision trees, inductive logic programming has been used,[22] and of course people have used neural nets[23] and genetic algorithms.[24] Some have been more successful than others, but none have so far matched the popularity of the statistical and knowledge-based methods. Research groups have compared different approaches to see how to make best use of them in combination, and significant projects continue in this area.[25]

A weakness with some of the research has been in approaching the problem from a mathematical, or information science, point of view rather than from the point of view of a chemist. A very simple example arose in a project I was involved in. A non-chemist used a general purpose data mining tool to look for links between chemical structure and toxicological activity. The mining tool automatically applied Occam's razor. That is, if several solutions were available, it selected the simplest one. I forget what activity was being modelled and what structures were in the training set, but that does not matter. Suppose that the end-point was skin sensitisation, all of the active molecules in the training set were acid chlorides and there were no examples of other chlorine-containing compounds. The system associated the presence of the chlorine atom with activity. It did not associate oxygen, or the carbonyl group, with activity because they were present in lots of inactive molecules in the training set. It discovered the pairing of a carbonyl group with the chlorine atom in every active molecule but, applying Occam's razor, it automatically ignored the carbonyl group since chlorine alone was enough to identify the active compounds. So it concluded that any compound containing a chlorine atom was likely to be a skin sensitiser. For the purposes only of classifying the contents of the training set the program was right, but a chemist would have seen at once that the simplification was likely to be a mistake in a chemical and biological context.

References

1. C. Hansch and T. Fujita, A Method for the Correlation of Biological Activity and Chemical Structure, *J. Am. Chem. Soc.*, 1964, **86**, 1616–1626.
2. T. Fujita, J. Isawa and C. Hansch, A New Substituent Constant, π, Derived from Partition Coefficient, *J. Am. Chem. Soc.*, 1964, **86**, 5175–5180.
3. ClogP is supplied by Biobyte Corporation, 201 West 4th. Street, #204, Claremont, CA 91711-4707, USA.
4. L. H. Hall, B. Mohney and L. B. Kier, The Electrotopological State: Structure Information at the Atomic Level for Molecular Graphs, *J. Am. Chem. Soc.*, 1991, **31**, 76–82.
5. R. W. Taft, Separation of Polar, Steric, and Resonance Effects, in *Steric Effects in Organic Chemistry*, ed. M. S. Newmann, John Wiley, New York, 1956, pp. 559–675.
6. M. J. Kamlet, J.-L. M. Abboud, M. H. Abraham and R. W. Taft, Linear Solvation Energy Relationships. 23. A Comprehensive Collection of the Solvatochromic Parameters, π^*, α, and β, and Some Methods for Simplifying the Generalised Solvatochromic Equation, *J. Org. Chem.*, 1983, **48**, 2877–2887.
7. L. Eriksson, E. Johansson, N. Kettaneh-Wold and S. Wold, in *Multi- and Megavariate Data Analysis*, Umetrics AB, Umeå, Sweden, 2001.
8. C. Hansch and A. Leo, in *Exploring QSAR: Fundamentals and Applications in Chemistry and Biology*, Am. Chem. Soc., Washington DC, USA, 1995.
9. D. Livingstone, in *Data Analysis for Chemists: Applications to QSAR and Chemical Product Design*, Oxford University Press, England, 1995.
10. K. Enslein and P. N. Craig, Carcinogenesis: a Predictive Structure-Activity Model, *J. Toxicol. Environ. Health*, 1982, **10**, 521–530.
11. K. Enslein, V. K. Gombar and B. W. Blake, Use of SAR in Computer-Assisted Prediction of Carcinogenicity and Mutagenicity of Chemicals by the TOPKAT Program, *Mutation Res.*, 1994, **305**, 47–62.
12. Accelrys Inc., 10188 Telesis Court, Suite 100, San Diego, CA 92121, USA. http//accelrys.com/
13. G. Klopman, Artificial Intelligence Approach to Structure-Activity Studies: Computer Automated Structure Evaluation of Biological Activity of Organic Molecules, *J. Am. Chem. Soc.*, 1984, **106**, 7315–7321.
14. G. Klopman, J. Ivanov, R. Saiakhov and S. Chakravarti, MC4PC – An Artificial Intelligence Approach to the Discovery of Structure Toxic

Activity Relationships (STAR), in *Predictive Toxicology*, ed. C. Helma, CRC Press, Boca Raton, 2005, pp. 423–457.

15. G. Klopman, S. K. Chakravarti, H. Zhu, J. M. Ivanov and R. D. Saiakhov, ESP: a Method to Predict Toxicity and Pharmacological Properties of Chemicals Using Multiple MCASE Databases, *J. Chem. Inf. Comput. Sci.*, 2004, **44**, 704–715.
16. P. N. Judson, Rule Induction for Systems Predicting Biological Activity, *J. Chem. Inf. Comput. Sci.*, 1994, **34**, 148–53.
17. R. E. Carhart, D. H. Smith and R. Venkataraghavan, Atom Pairs as Molecular Features in Structure-Activity Studies: Definition and Applications, *J. Chem. Inf. Comput. Sci.*, 1985, **25**, 64–73.
18. R. Kühne, F. Kleint, R. -U. Ebert and G. Schüürmann, Calculation of Compound Properties Using Experimental Data from Sufficiently Similar Chemicals, in *Software Development in Chemistry 10*, ed. J. Gasteiger, Gesellschaft Deutscher Chemiker, Frankfurt, Germany, 1996, pp. 125–134.
19. R. Kühne, R. -U. Ebert and G. Schüürmann, Estimation of Compartmental Half-Lives of Organic Compounds–Structural Similarity vs. EPI-Suite, *QSAR Comb. Sci.*, 2007, **26**, 542–549.
20. LeadScope Predictive Data Miner comes from Leadscope Inc., 1393 Dublin Road, Columbus, Ohio 43215, USA.
21. J. R. Quinlan, Induction of Decision Trees, *Machine Learning*, 1986, **1**, 81–106.
22. R. D. King and A. Srinivasan, Prediction of Rodent Carcinogenicity Bioassays from Molecular Structure Using Inductive Logic Programming, *Environ. Health, Perspect.*, 1996, **104**, 1031–1040.
23. M. Vracko, V. Bandelj, P. Barbieri, E. Benfenati, Q. Chaudhry, M. Cronin, J. Devillers, A. Gallegos, G. Gini, P. Gramatica, C. Helma, D. Neagu, T. Netzeva, M. Pavan, G. Patlevicz, M. Randic, I. Tsakovska and A. Worth, Validation of Counter Propagation Neural Network Models for Predictive Toxicology According to the OECD Principles. A Case Study, *SAR QSAR Environ. Res.*, 2006, **17**, 265–284.
24. F. V. Buontempo, X. Z. Wang, M. Mwense, N. Horan, A. Young and D. Osborn, Genetic Programming for the Induction of Decision Trees to Model Ecotoxicity Data, *J. Chem. Inf. Model.*, 2005, **45**, 904–912.
25. C. Helma, T. Cramer, S. Cramer and L. De Raedt, Data Mining and Machine Learning Techniques for the Identification of Mutagenicity Inducing Substructures and Structure Activity Relationships of Non-congeneric Compounds, *J. Chem. Inf. Comput. Sci.*, 2004, **44**, 1402–1411.

CHAPTER 12

The 2D–3D Debate

Forget about chemistry for the moment and consider Figure 12.1. It is a picture of a hexagon. It lies in the plane of the paper and can be fully described in terms of the locations of its vertices, using two-dimensional Cartesian coordinates (x, y). Reconsider Figure 12.1 as a chemical diagram. Now it is cyclohexane. The picture can still be fully described using 2D coordinates but the molecule it represents is not planar and if you want to define its shape in 3D space you need 3D coordinates (x, y, z). A perspective picture, giving an illusion of its shape to the human eye closer to reality, might be the one in Figure 12.2. In the first eleven chapters of this book, describing a variety of computer systems dealing with chemistry, there have been some passing references to stereochemistry, implying an awareness of the three-dimensionality of chemical structures, but no mention has been made of 3D coordinates. Is that not a serious oversight?

Molecular modelling, the study of the interactions between molecules in 3D, has become an entire sub-discipline. When I was a student, a lecturer was able to tell us that, for the first time, computer power was sufficient to allow calculation of the size and shape of the hydrogen molecule, H_2, from first principles. By now the dimensions of much more complicated structures can be calculated from first principles. So-called semi-empirical methods and molecular mechanics make it possible to build 3D structures without having to do such heavy computation, and structures for compounds as large as proteins are available from X-ray crystallography. Chemists can look at interactions in 3D between molecules – for example, the binding of a pharmaceutical or toxic compound to its site of action in a protein.

In cases where the members of a group of compounds are believed all to bind to the same site but the structure of the site itself is not known, the requirements for binding can be worked out by superimposing the structures of the active compounds onto each other. You might find, for example, that you can draw the same triangle between the location of a hydrogen bond donor, a second

RSC Theoretical and Computational Chemistry Series No. 1
Knowledge-based Expert Systems in Chemistry: Not Counting on Computers
By Philip Judson
© Philip Judson 2009
Published by the Royal Society of Chemistry, www.rsc.org

Figure 12.1   A hexagon or cyclohexane.

Figure 12.2   Cyclohexane.

hydrogen bond donor, and the centre of an aromatic ring in all the structures. If there are four centres in common between the active molecules they will form the apices of a tetrahedron. In either case, the distances between the centres will be distances in 3D space. If the biological activity of the compounds is a pharmacological one (which is most often the case because it is primarily in pharmaceutical research that molecular modelling is used) these representations are called 3D pharmacophores, or just pharmacophores. They differ from pharmacophores as described in Chapter 10, those being defined by connectivities between atoms and bonds, not by 3D distances. In discussions with other researchers it is important to make it clear which kind of pharmacophore you are talking about.

Most molecules of interest contain single bonds, and rotation about those bonds allows their structures to take on many shapes, or "conformations". A structure arranged in a particular conformation is generally referred to as a "conformer". Some conformations will be energetically less favourable than others. For example, as you rotate a carbon–carbon single bond there will be three positions, 120° apart, in which the substituents at opposite ends of the bond are pushed up against each other (the "eclipsed" conformations) and three, rotated 60° from the first set, in which there is minimal contact between them (the "staggered" conformations).

For a pharmaceutical molecule to bind to a biological site – for example the active site in an enzyme – it needs to adopt a particular conformation. Strength of binding depends on the energy saving associated with binding. So to predict how strongly a novel structure will bind to a site, you need to estimate the binding energy and compare it with the energy needed for the structure and the site it binds to to adopt the necessary conformations, which may not be the ones preferred by the structure and site when separated. As you rotate bonds, interactions between substituents will increase and decrease, and the peaks and troughs will differ in size (see Figure 12.3). Some minima that are higher than the global minimum, may still be low enough for binding if the binding energy is sufficient, and so you need to consider all the appropriate low energy conformations. For each conformation, some movement will be possible without too great an energy cost. So the distances between the apices of a 3D pharmacophore will not be precise; each will fall within a range.

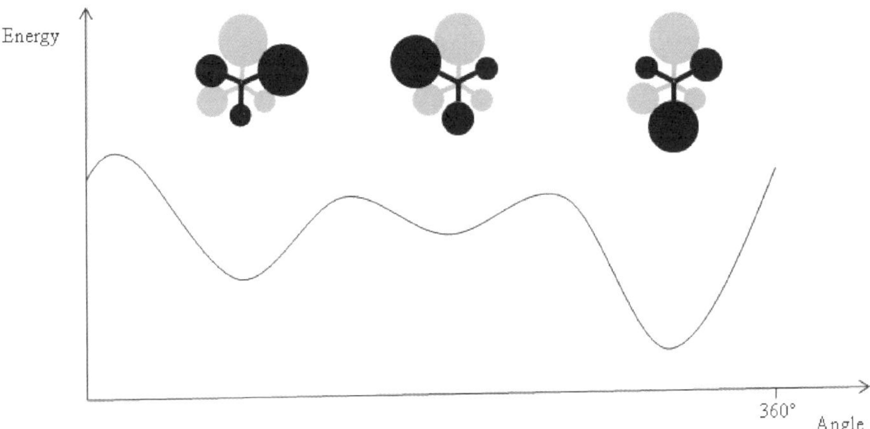

Figure 12.3   Changing energy as a bond rotates.

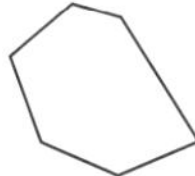

Figure 12.4   A valid, if untidy, representation of cyclohexane.

The programs described in the earlier chapters of this book, and the methods they use, have often been casually termed "2D", because they communicate through two dimensional chemical diagrams. That is an incorrect description. A chemical diagram is not just a two dimensional drawing. It is a graph. Indeed, there is no significance in the 2D layout of the diagram, beyond a consideration of aesthetics, as long as the connectivity is right. The information the diagram conveys is what kinds of atoms are joined by what kinds of bonds. Figure 12.1 and Figure 12.2 are both recognisable to a chemist as representations of cyclohexane, and so would be Figure 12.4, even if its style is somewhat eccentric. Because of the directional nature of chemical bonding, a lot is implied about 3D shape and distances by a chemical graph. Alerts, pharmacophores, *et cetera*, defined by sub-structural fragments are therefore sometimes said to be "2.5D", rather than "2D". They might alternatively be called topological, a term that I have preferred, but there is a problem with it because to many chemical information scientists and molecular modellers "topological descriptors" are numerically encoded forms of topological information generated for use in mathematical equations. So in this book I will use the term "2.5D" even though it is not entirely satisfactory.

Using 2.5D pharmacophores and toxicophores is not as different from using 3D ones as you might at first suppose. Imagine that a 2.5D pharmacophore includes two atoms separated by three single bonds. According to my calculations, the distance between the two atoms, assuming the bonds to be carbon–carbon bonds will be roughly in the range 2.5–3.8 Å. That information is implicit whether or not you calculate it – the distance between the relevant atoms in any substructure that matches the 2.5D pharmacophore will be in that range. So, although the breadths of ranges may differ, this is equivalent to the use of distance ranges in 3D modelling.

There is a further matter to take into consideration when using 3D pharmacophores and toxicophores: molecules are not static; they are in constant motion – vibrational as well as translational and rotational. Docking methods – whether automated or manual – allow molecules to move into and out of the binding sites (translational motion) and to tumble as they do so, and they allow internal rotations to take place so that the molecules can adopt the right shape to bind, but the finalised model of a molecule in its bound state assumes that the molecule is more or less stationary. In reality vibration and internal rotation of parts of a structure can make big differences to binding strength. Molecular dynamics can help in some cases, but for the kinds of interactions that people typically want to model the computational demands are too great.

The limitations on how precise 3D modelling can be on the one hand – some inherent, some practical – and the fact that 2.5D modelling implicitly takes more into account than you might at first have thought, mean that there is often not much difference in the predictive usefulness of the two methods. The references at the end of this chapter provide three examples of the dozens of papers have been published comparing a variety of 2.5D and 3D descriptors for the prediction of biological activity.[1–3] In some cases 3D models have been found to have advantages, but 2.5D models are often equally effective. Sometimes 2.5D models even work better than 3D ones – 3D models may be too restrictive with regard to conformational flexibility or it may be too difficult to determine which conformers should be favoured. In addition, 2.5D models have computational advantages over 3D models: calculations based on graph theory are much less demanding on processor power and memory capacity than are the calculations about electron distribution, energy levels, and so on required in molecular modelling, and, for the reasons discussed earlier in this chapter, many calculations for many conformers need to be done in 3D modelling, whereas the problem is side-stepped in 2.5D modelling.

The applications discussed in this book, both in earlier and succeeding chapters, use 2.5D models. Some would be capable in principle of using 3D information, but knowledge base developers working on Derek for Windows, for example, (the subject of Chapter 13) have not to date found cases where they deem it to be necessary. Perhaps 3D methods are more precise than 2.5D methods, but being precise may be a hindrance if you are trying to model events that are inherently imprecise. An underlying theme in most of the rest of this book, is the development of ways to work with things that are imprecise and uncertain without trying to use precise theories and models for them. It is an attitude to life with which organic synthesis chemists will be comfortable – and biologists, too. People with different scientific backgrounds may be less so, but accepting that imprecise things are imprecise and treating them that way is more, not less, scientific than willing them into being precise.

References

1. R. D. Brown and Y. C. Martin, The Information Content of 2D and 3D Structural Descriptors Relevant to Ligand-Receptor Binding, *J. Chem. Inf. Comput. Sci.*, 1997, **37**, 1–9.
2. S. C. Basak, B. D. Gute and G. D. Grunwald, Relative Effectiveness of Topological, Geometrical, and Quantum Chemical Parameters in Estimating Mutagenicity of Chemicals, in *Quantitative Structure–Activity Relationships in Environmental Sciences VII*, ed. F. Chen and G. Schüürmann, SETAC Press, Pensacola, FL, 1998, pp. 245–261.
3. S. C. Basak, R. Natarajan, D. Mills, D. M. Hawkins and J. J. Kraker, Quantitative Structure-Activity Relationship Modeling of Insect Juvenile Hormone Activity of 2,4-Dienoates Using Computed Molecular Descriptors, *SAR QSAR Environ. Res.*, 2005, **16**, 1–26.

Making Use of Reasoning: Derek for Windows

13.1 Moving on from Just Recognising Alerts in Structures

The DEREK program warned a user about the potential toxicity of a query structure on the grounds that it contained an alert – a substructural feature that toxicologists believe to interact with biological systems leading to toxicity by a particular mechanism. DEREK did not usually indicate how likely a toxic effect would be, or how severe it would be. A special case was the neural toxicity arising from acetylcholinesterase inhibition mentioned in Chapter 9, for which there was a warning that activity was likely to be high if the substructure triggering the organophosphate alert contained a substituted amino group four bonds distant from the phosphorus atom of the phosphate group. Apart from that, DEREK simply recognised an alert, highlighted it on the computer screen, and stated what toxicological end-point was associated with it.

Statistical QSAR systems give numerical estimates of toxic potency and programs such as HazardExpert[1] give a numerical estimate of the probability that a compound will be toxic. However, there are difficulties with numerical approaches. The question of whether it is sound science to apply the rules of chance to prediction of toxicity is raised in Chapter 10, and practical problems with using standard methods of data analysis are mentioned in Chapter 11.

The panel in a recent BBC radio programme[2] were presented with the following problem. A stage magician takes one card from a pack and puts it into a box. He invites the audience to name a card and someone shouts out "Ace of hearts". When he displays to the audience the card from the box, what are the chances that it will be the ace of hearts? One panel member started to say that

RSC Theoretical and Computational Chemistry Series No. 1

Knowledge-based Expert Systems in Chemistry: Not Counting on Computers

By Philip Judson

© Philip Judson 2009

Published by the Royal Society of Chemistry, www.rsc.org

most of the story was a distraction, and the probability that the card would be the ace of hearts was $1/52 \cong 0.019$. The question master asked why she was so sure the rest of the story was a distraction and she realised, of course, that this question was about a performance by a magician. Assuming he was good at his job, the probability that the card he held up would be the ace of hearts was close to 1.0. Her colleague joined in to remark that in some performances the card the magician holds up is not the one chosen by the audience. The magician might feign disappointment and then fan out the rest of the pack to reveal that every other card in it was the ace of hearts. The panel's conclusion was that the probability that the card from the box would be the ace of hearts was less than 1.0 but a lot more than 0.019. The laws of chance are context dependent.

Whatever your views about using numerical methods for toxicity prediction, perhaps the biggest practical problem is that numerical methods require numerical inputs, and reliable ones are often not available. On the other hand, without numbers, what can you do? Could advice from a program like DEREK include a soundly-based assessment of how likely it would be that activity would be expressed, without having to use numerical probability or statistics?

It is very often the case – perhaps almost always – that whether a compound containing an alert is active or not depends upon its fat–water partition properties, more conveniently related to its octanol–water partition coefficient, "K_{ow}" or "log P", which can be measured or predicted fairly reliably.[3] There are at least two reasons for the influence of partition coefficient on toxicity, either or both of which may apply in a given case. To exert its toxic effect, a compound has to reach its site of action. Whether it enters the body by ingestion, through the skin, or by inhalation, the absorption process depends upon the partition properties of a compound and, thereafter, so do its progress around the body, its penetration into the cells where the toxicological action takes place, and its success in crossing membranes to the site of action. If the compound acts by a lock-and-key mechanism, requiring it to bind to a site on a protein, the tightness of binding depends on partition properties – lipophilic molecules, or parts of molecules, will bind to lipophilic surfaces of the protein – hydophilic molecules, or parts of molecules, to hydrophilic surfaces.

Modifications were made to DEREK so that it could ask the user whether the query structure was likely to have a high, medium, or low log P, and/or could use a log P value provided in an input file, which was classed as high, medium, or low according to cut-off values set by knowledge base writers. Rules were written in CHMTRN such as that "if the log P is high, then strong activity is likely", "if the log P is medium or low then activity is likely to be absent or weak". As these examples reveal, there was some ambiguity about whether likelihood of activity or potency was being predicted. In practice, this may not matter much, since the two are confused in the definition of toxicity anyway: a compound is considered to be toxic if its activity is above a certain level; not every active compound is active enough to be classified as toxic; while the likelihood of activity may be a pure concept, the likelihood of toxicity, as normally defined, depends also on potency. However, there was no overall scheme behind the design of DEREK for bringing together supporting or conflicting evidence about likelihood of activity or potency. So when John Fox, then head of the advanced computation laboratory at Imperial Cancer Research Fund, suggested collaborating on the use of reasoning methods that were being developed by his laboratory to assess the likelihood of toxicological outcomes qualitatively it was of great interest to us at Lhasa Limited.

13.2 The Logic of Argumentation

The reasoning being developed by the research group at Imperial Cancer Research Fund depended upon the Logic of Argumentation (LA).[4,5] Models for computer reasoning based on LA had been described before and used in prototype systems for supporting medical diagnosis.[6,7] Having been a member of a committee to review a project carried out for the Ministry of Agriculture, Fisheries, and Food by Lhasa Limited, John Fox had seen DEREK and recognised the potential for working together on a system to predict carcinogenicity. In a collaboration, his team could provide expertise in LA and the broader issues of reasoning under uncertainty, and Lhasa Limited could provide expertise in the handling of chemical structures by computer systems. Logic Programming Associates, a software company specialising in Prolog compilers and related tools for logic-based software development, and psychologists at City University, with an interest in how people perceive and communicate about risk, also joined the collaboration. The Department of Trade and Industry granted funding for the project, which was given the name "StAR", derived from "Standardised Argumentation Report", a term used by John Fox for the formalised presentation of the reasoning behind a decision based on LA.

The equations for calculating numerical probabilities, in the range 0 to 1, originally came out of an interest in the mathematics of chance. This kind of probability has been termed "stochastic" probability. Psychological research shows that humans reach decisions through a process of reasoning.[8] This may seem obvious, but the experiments provide the scientific confirmation. When a human brain assesses how likely it is that something will happen, the judgement is based on experience (which may include things that have been learned from other people) and not on a numerical probability calculation. This notion of "epistemic" probability, probability based on past experience, pre-dated the development of the mathematics of stochastic probability by many centuries.

If I ask you whether the traffic will be bad on the way into town at 8.45 tomorrow morning you will not do calculations using data on people's home and work addresses, start times at work, and intentions to go to work or take the day off; your answer will be based on what the traffic is usually like at that time of day. That does not mean that your assessment of how likely it is that the traffic will be bad is necessarily superficial. You will take into account the time of year, whether there is a national holiday reducing the number of people travelling to work, or an international cricket match on that day at a stadium along the same route, and so on. You can give surprisingly reliable guidance without the need to do any arithmetic. There are obvious hazards with basing decisions on past experience – people, individually and collectively, regularly show seemingly irrational bias – but broadly speaking our successful survival and evolution show that the method has stood us in good stead.

LA assesses the arguments for and against a proposition in order to reach a conclusion, rather in the way that a court of law operates. Evidence that the knave of hearts stole some tarts might be that, shortly after the disappearance of the said tarts from the kitchen, he was found to have jam on his thumb. Arguments against his having stolen the tarts might be that he denied it absolutely, and that he had a friend willing to swear they were both somewhere else at the time that the tarts disappeared. The case for the prosecution might be demolished by evidence from an expert witness for the defence that the jam on the knave's thumb was raspberry jam, whereas it had been strawberry tarts that had gone missing from the kitchen. This illustrates a feature of reasoning that needs to be correctly interpreted. The fact that the knave's thumb only had raspberry jam on it provides no support for the claim that he is innocent; it only means that the presence of jam provides no evidence for his guilt. Such a counter argument is termed an undercutting argument. If there is no other evidence to show he is guilty, the case will collapse, but because it is "not proven" not because he is assuredly "not guilty". If his alibi stands, of course the case is clear cut – he could not have been guilty.

What the output of an LA model should be – the likelihood that something will happen, belief that it will, or confidence that it will – is as much discussed as are the meanings of the words, "likelihood", "belief" and "confidence". There is clearly a difference between the objective concept of how likely it is that something will happen, and the more subjective ones of how strongly you believe it will happen or how confident you are about a prediction. For the purposes of this book, I use the term "likelihood", which is also used in the program, Derek for Windows, but be aware that there is room to disagree about whether this is the right word to use. When you move from the theoretical to the practical world, dividing lines tend to get untidy. There is a fine line between "how likely something is" and "how likely something appears to be", and that looks more like a definition of "belief". Using the word "likelihood" is also not ideal, in that for statisticians it has a specific, different meaning from the one intended here, but other words that come naturally to mind – probability, belief and confidence – also already have specialised meanings.

The reasoning process in LA is built around arguments of the form

    If <grounds> are <threshold> then <proposition> is <force>

    where: <grounds> are the evidence on which an argument is based;

        <threshold> is the minimum strength of evidence for which the argument holds:

        <proposition> is what the argument predicts;

        <force> is the strength of the argument.

If this appears obscure, an illustration should help:

"If late_in_the_day is certain then inadequate_light_for_cricket is probable".

The grounds of this argument are "late_in_the_day" and the threshold is "certain". To paraphrase, the first part of the statement is saying "if you are sure that it is late in the day . . . ". The force of the argument in this example is "probable". Precisely what "probable" means does not matter for the moment. The point is that it is not "certain" that the light will be inadequate, only probable. It might be an unusually clear evening and the match might be being played close to midsummer. The purpose of the threshold ("certain" in this case) is not obvious from this example. More generally in a reasoning environment you might want to say things like "If adequate_takings_at_the_gate are *doubted* then cancellation_of_the_match is probable": the match organisers are not likely to need proof that they will fail to cover costs before they decide to cancel if finances are tight – being in doubt will be enough.

In a traditional system of logic the grounds and conclusions of arguments are either true or false. So if something is not true it must be false. That assumption is not valid if there is uncertainty in a prediction. Take the statement "If it has rained the pavement will be wet". If it is true that it has rained, then we can be sure that the pavement will be wet. But if it has not rained, the pavement might be dry, but it might be wet for all sorts of other reasons. In LA arguments for and against propositions operate independently; the failure of an argument for something provides no evidence against it, and *vice versa*. The arguments for a proposition and those against it are first separately aggregated, and then the overall case for is weighed against the overall case against to reach a conclusion. There are many opportunities for debate about how the aggregation and the final resolution should be done[5] and in some cases there may be room for flexibility, depending on the context in which LA is being used. Just the basics are discussed here.

Figure 13.1 is a graphical representation of a small reasoning tree showing the different ways in which propositions can interact. 'A' is the grounds of an argument leading to proposition 'C'. 'a' is the force of the argument, and 'a′' is its threshold. So if the likelihood of 'A' is at least equal to 'a′' then the

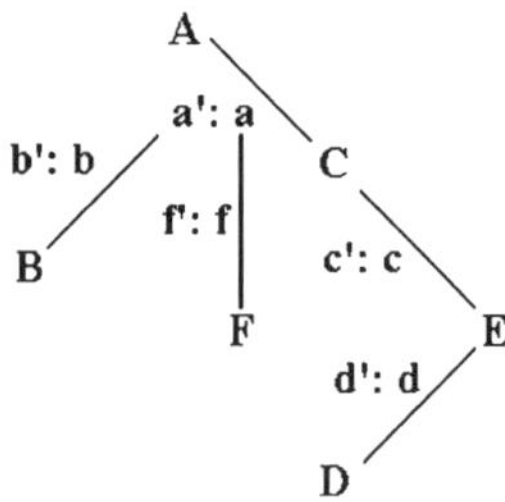

Figure 13.1 A simple reasoning tree.

likelihood of 'C' will be assigned the value 'a' on the basis of this argument. 'C' in its turn is the grounds of a second argument leading to proposition 'E'. Another argument relating to 'E' is based on grounds 'D'. Also illustrated in the figure is the idea that how likely something is on the basis of one argument might depend on other arguments – the value of 'a' may depend on the argument with grounds 'F'. Even the threshold above which an argument comes into effect may depend on other arguments – illustrated by the argument with grounds 'B' that determines the value of 'a′'.

Well, that is all splendidly obscure. Perhaps a couple of examples will help. Let us start with three simple propositions:

$$\text{If 'overcast\_sky' is 'true' then 'rain' is 'probable'.} \qquad (1)$$

$$\text{If 'rain' is 'true' then 'we\_will\_get\_wet' is 'probable'.} \qquad (2)$$

$$\text{If 'under\_an\_umbrella' is 'true' then 'we\_will\_get\_wet' is 'improbable'.} \qquad (3)$$

These propositions can be related to the graph in Figure 13.1 if you leave out the arguments in the graph based on 'B' and 'F'. 'A' is 'overcast_sky', 'a′' is 'true', and 'a' is 'probable'; 'C' is 'rain', 'c′' is 'true', and 'c' is 'probable'; 'D' is under_an_umbrella', 'd′' is 'true', and 'd' is 'improbable'; 'E' is 'we_will_ get_wet'. Consider a dismal day when overcast_sky is true. It follows from argument (1), that rain is probable. But if rain is probable, then according to argument (2) we_will_get_wet is probable. If neither we nor any of our friends take umbrellas, then under_an_umbrella will regrettably be impossible. Remembering the rule that an argument against a proposition makes no contribution to the case for it, the output from argument (3) is open. So the bad news is that it is probable we will get wet.

We will come back to the umbrella in a moment, but first, to complete the story, note that if there were a clear, blue sky, overcast_sky would be false and the output from argument (1) would be open. In its turn, the case for rain being open would mean that the output from argument (2) would be open as well. Whether we might get wet would remain open. You might wonder why the conclusion should not be that we_will_get_wet is false. The rule that information against a proposition makes no contribution to the case for it has a logical basis: the circumstance of rain not falling does not contribute to keeping you dry; it fails to get you wet which is not the same thing. If the neighbour's children assault you with water pistols the absence of rainfall will do nothing to save you from the consequences.

Let us assume for the moment that by "probable" we mean more likely than not. What if the likelihood of an overcast sky, on the basis of the weather forecast, is "just about conceivable"? Argument (1) does not fail but it is weakened. Rain is no longer probable. It can only be "just about conceivable". Applying argument (2) leads us similarly to conclude that getting wet is "just about conceivable". This illustrates one of the rules for propagating arguments along a chain of reasoning, namely that the likelihood attached to the proposition of an argument (provided that it does not fail) is the smaller of the force of the argument and the magnitude of likelihood of the grounds of the argument. I use the clumsy "magnitude of likelihood of the grounds" for a reason. Suppose that instead of argument (1) we had the argument "If clear_sky is false then rain is probable". If it were "just about conceivable" that clear_sky was *false* this would lead to the conclusion that it was "just about conceivable" that rain would be *true*.

Having previously been unwise enough to go out without an umbrella on a day when overcast_sky was true, let us take one with us this time and use it when the need arises. Under_an_umbrella is now true and so according to argument (3) we_will_get_wet is improbable (in case you are wondering, I did not make the force for argument (3) "impossible", since there are lots of ways you still might get wet with only the limited protection of an umbrella). But according to argument (2) we_will_get_wet is probable. Supposing that the terms "probable" and "improbable" carry equal weight for and against, we have to conclude that whether we will get wet is equivocal. Reasoning models that implement LA define rules for resolving conflicting arguments for and against a proposition, one of which could be the resolution of "probable" and "improbable" into "equivocal".

If you have a set of arguments of different forces for or against a proposition but there is one that *proves* the proposition to be true or false, then clearly that conclusion must prevail. It does not matter how much circumstantial evidence you have to suggest that the knave of hearts stole the tarts, if there is incontrovertible proof of his alibi then he cannot have done it (at least, not in person). Simultaneous proof that something is true and that it is false – contradiction – is a theoretical concept that needs to be supported in a complete LA model because real-world cases will arise in which there is apparent contradiction. At least one of the pieces of evidence must be wrong if the laws of common sense are to be trusted, but the reasoning model needs to be able to cope with the situation.

If you have several arguments for a proposition, is the proposition more likely than it would have been were there only one argument for it? The answer turns out to be dependent both on the definitions you use for measures of likelihood and the context in which you use your model. An example where the proposition is more likely if there are multiple arguments for it, is familiar to anyone who has learned the laws of probability. If the numerical probability of being hit by a falling brick is x and the probability of being hit by a falling meteorite is y, then the probability of being hit by something is greater than x or y. But suppose you have a hierarchy of terms in a reasoning model defined as follows: "certain" means there is proof that something is true; "probable" means that there is at least one argument for something and there are no arguments against it; "plausible" means there are arguments both for and against something but the balance of the arguments is for. It does not matter how many arguments you have that something is probable, they will not constitute proof that it is true. It does not matter how many arguments you have that something is plausible, they cannot make it probable, as defined, because being plausible means there are arguments against and so one of the criteria for "probable" cannot be met. A given LA model will specify which way multiple arguments for (or against) a proposition are aggregated, but the most favoured one, and the one that appears to be most useful in practice, is the one that says that the force for a proposition is simply the greatest of those of the set of arguments for it and the force against a proposition is the greatest of those of the set of arguments against it.

To return again to our sortie under uncertain skies, a different set of arguments illustrates the use of undercutting and of changing the force of an argument according to circumstance, instead of depending on conflict between arguments. Argument (4) is the same as argument (1), but it is convenient to repeat it here.

$$\text{If overcast\_sky is true then rain is probable} \tag{4}$$

$$\text{If rain is true then we\_will\_get\_wet is} \\ \text{<exposure\_to\_the\_elements>} \tag{5}$$

$$\text{If under\_an\_umbrella is false then} \\ \text{<exposure\_to\_the\_elements> is true} \tag{6}$$

$$\text{If under\_an\_umbrella is true then} \\ \text{<exposure\_to\_the\_elements> is equivocal} \tag{7}$$

In this set of arguments <exposure_to_ the_elements> is a variable. Seeing an overcast sky, we note that according to argument (4) rain is probable. 'Probable' is a measure of likelihood in favour of something and so argument (5) applies. The output from argument (5) should be either 'probable' or <exposure_to_ the_elements>, whichever is the weaker. But what is <exposure_to_the_elements>? Suppose first that under_an_umbrella is false, then the output from argument (7) is 'open' and argument (6) tells us that 'exposure_to_the_elements' is 'true'. 'Probable' being weaker than 'true' the output from argument (5) becomes 'probable': it is probable that we_will_get_wet. Now suppose that we are sheltering under an umbrella; 'under_an_umbrella' is true. The output from argument (6) is 'open', but from argument (7) we find that <exposure_to_ the_elements> is equivocal. So the output from argument (5) is the weaker of 'equivocal' and 'true'. We conclude that we_will_get_wet is equivocal.

The first model used for illustration above, using arguments (1) to (3), is based on the notions that rain presents a threat of getting wet whatever the circumstances and that possession of an umbrella is a defence against getting wet from whatever cause, although not one hundred per cent effective. The second model is based on the notions that rain threatens to make you wet only if you are exposed to it, and that an umbrella provides modest protection against exposure. In my view, the second model is the better representation of reality. The answers that you get from the two models about rain and umbrellas are the same, but the logical courses to them are different. "Yes but," you denounce, "They are the same, because you have chosen different values of force for the arguments in the two models to make them the same!" It is true. For the purposes of illustration, I made sure the answers were the same. None of the forces I used were based on reality anyway. To do the job properly, I should either have assessed how well an umbrella works in practice – given the complications of gusty winds, passing buses hitting puddles, and so on – or, better, written a lot of arguments about those other factors to create a more complete model. Choosing the right model for a real application requires a lot of thought, and, of course, the values you choose for the forces of arguments should be based on evidence, not on what you want the answers to be.

I suspect that some of the things presented above are harder to describe than they are to understand, so if you cannot get your head round what you have just been reading, blame the writer. The same ideas, together with some issues not covered here, are presented in ways that might or might not suit you better in two of the references at the end of this chapter.[3,9]

13.3   Choosing Levels of Likelihood for a System Based on LA

Toxicologists interested in the StAR project told us that they would like the computer system to indicate how likely toxicity would be, but only in broad terms. They did not want spurious precision such as being told that there was an 11.49% likelihood of toxicity, when all that could really be said was that "toxicity might be seen, but it was not all that likely". They felt that we should use only a few levels of likelihood. But how should we define those levels, and how should we name or label them? It was agreed that for the application being developed in the StAR project we would define five levels of uncertainty and that they would interact with each other according to consistent rules, in order to provide a reliable model. The definitions that were decided upon can be described as follows:

> there is at least one strong argument that the proposition is true and there are no arguments against it;
>
> there are no strong arguments for the proposition, or there are arguments both ways, but the weight of evidence supports the proposition;
>
> there are equally strong arguments for and against;
>
> there are no strong arguments against the proposition, or there are arguments both ways, but the weight of evidence opposes the proposition;
>
> there is at least one strong argument that the proposition is false and there are no arguments to support it.

Four other states needed to be covered, the first two for cases where there is certainty rather than uncertainty, and the second two for cases where no meaningful conclusion can be reached:

there is proof that the proposition is true;

there is proof that the proposition is false;

the proposition is apparently proved to be both true and false;

there is no evidence that supports or opposes the proposition.

Numbers used as labels do not tell you anything in the absence of advice about the scale they represent. If I tell you that the likelihood that a chemical will be toxic is 17, it means nothing. If, on the other hand, I tell you that it is *very likely* that a chemical will be toxic, it is pretty clear to you that putting the chemical on your dinner would not be a smart decision. So there appears to be a case for using words to name levels of likelihood in a prediction system. However, I do not know whether what you think "very likely" means, is what I think it means. The psychologists in the StAR project conducted research in which subjects were given cards each of which had one of the above descriptions of a state of certainty or uncertainty, and asked to choose words to describe them by selecting from a second set of cards each bearing one word. The kinds of words offered were "certain", "possible", "doubted", "dubious", "incontrovertible", "plausible", "improbable", "tenuous". In parallel experiments, different subjects were asked to do the reverse – to choose the description that best fitted each word.

It became apparent that the one word you should avoid at all costs, especially when talking about risks and threats to an apprehensive audience, is "possible" – the word most often on the tongues of expert scientists in national debates about hot political issues! The trouble with "possible" is that it means at least three things. You can be sure that many of your listeners will be picking up the wrong message. "So, Professor, this study appears to indicate that chlorination of tap water causes ingrowing toenails in susceptible minorities. Do you think it is true?" "It is possible on the basis of the data presented, but I would want much clearer evidence before I thought there was reason for concern." The Professor intends "possible" to mean that it cannot be ruled out – perhaps his thinking as a scientist is only that negative proof is lacking. Many listeners will have taken "possible" to have its second meaning – that it can be ruled in. They were probably too preoccupied with worrying about the high incidence of ingrowing toenails in their family to pay attention to the rest of his sentence. And what about another meaning of possible? If the answer to "Is it possible get to the bus station along this street?" is "yes" and you go along the street you will, *with certainty*, get to the bus station. That is a long way from merely being told that you cannot rule out getting there!

. . . If it is not too late to prevent a national panic, there is no link between chlorination and ingrowing toenails – I made it up.

The subjects used for the research to find suitable words were students whose first language was English. It would not be surprising if research using people

Table 13.1   Names and definitions of likelihood levels in Derek for Windows.

Likelihood	Definition
certain	There is proof that the proposition is true
probable	There is at least one strong argument that the proposition is true and there are no arguments against it
plausible	The weight of evidence supports the proposition
equivocal	There is an equal weight of evidence for and against the proposition
doubted	The weight of evidence opposes the proposition
improbable	There is at least one strong argument that the proposition is false and there are no arguments for it
impossible	There is proof that the proposition is false
open	There is no evidence that supports or opposes the proposition
contradicted	There is proof that the proposition is true and proof that it is false

whose first languages were not English and people from widely different educational backgrounds, came to different conclusions. However, the words that seemed most appropriate on the basis of the research were adopted for the StAR project, and subsequently used in Derek for Windows.[9] They are listed in Table 13.1. In practice, as knowledge increases, having only five levels of uncertainty becomes too limiting. Finding definitions for new levels that have provable relationships with the existing ones is not trivial, but there are some candidates. The bigger problem is to find words to label them with, without adopting a plethora of terms that would be obscure, even to a user whose first language was English. It seems likely that the use of words to describe levels of uncertainty, will have to be replaced by numbers, letters, some kind of symbol such as a column whose height represents a degree of likelihood, or a system of colour coding.

13.4   Derek for Windows

The StAR project produced a demonstrator system for advising on potential carcinogenicity.[10–12] Because of the immediate availability of knowledge, the first commercially-useful application using the technology to be released by Lhasa Limited predicted skin sensitisation (Unilever had funded the development of a skin sensitisation knowledge base for DEREK).[13–15] Soon after that a full reasoning-based product for predicting chemical toxicity more generally was released as Derek for Windows,[9,16] containing all the knowledge that DEREK contained. Derek for Windows underwent, and continues to undergo, further development and quickly displaced DEREK. All of the examples in this chapter are taken from Derek for Windows version 10.0.2.[17]

Figures 13.2 to 13.5 show how reasoning can change predictions made by Derek for Windows. For these examples, Derek for Windows was set to predict only for the skin sensitisation end point and with species limited to humans and mice.

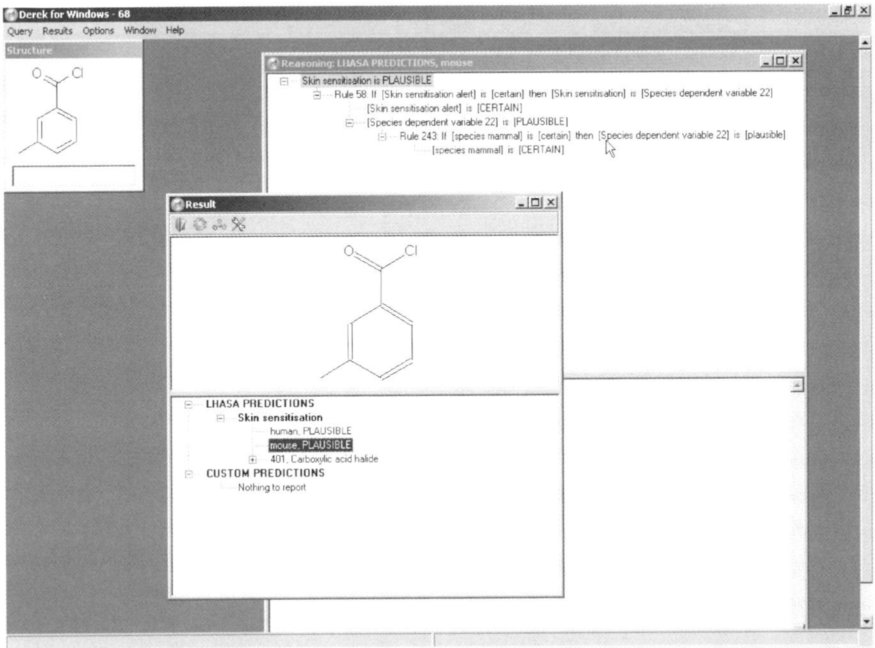

Figure 13.2   Skin sensitisation prediction for 3-methylbenzoyl chloride by Derek for Windows.

Figure 13.2 shows the prediction for 3-methylbenzoyl chloride. Skin sensitisation is predicted to be plausible in both humans and mice because alert number 401, associating skin sensitisation with carboxylic acid halides, was triggered by the query. The window displaying the reasoning shows only rules that have led to the reported prediction. The user would be overwhelmed with irrelevant information if rules that made no contribution to the conclusion were displayed. According to rule 58, "If [skin sensitisation alert] is certain then [skin sensitisation] is [species dependent variable 22]". [Skin sensitisation alert] is certain because skin sensitisation alert number 401 has been triggered. According to rule 243, "If [species mammal] is certain then [species dependent variable 22] is [plausible]". The program knows, from a taxonomy not shown in this figure, that human and mouse are both mammals and so in both cases [species mammal] is certain. So [species dependent variable 22] is plausible, which makes skin sensitisation plausible for both species according to rule 58.

Figure 13.3 shows the result window for benzoyl chloride with the triggering alert (number 401) highlighted. For this compound, skin sensitisation in humans is predicted to be probable and in the mouse it is stated to be certain. Figure 13.4 shows the reasoning that led to the conclusion for humans. There is no need here to go through the details of how rules 5 and 261 work together, since the process is the same as the one described above for rules 58 and 243. [Known local lymph node assay positive in mouse] is certain, *i.e.* benzoyl

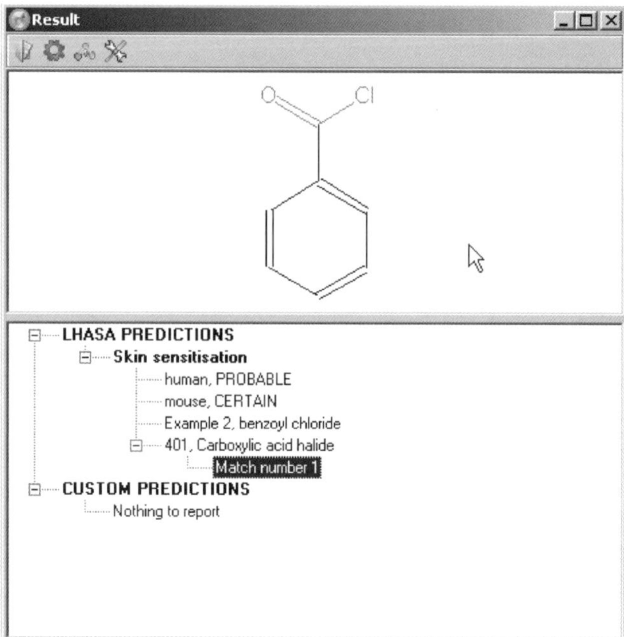

Figure 13.3   Skin sensitisation prediction for benzoyl chloride by Derek for Windows.

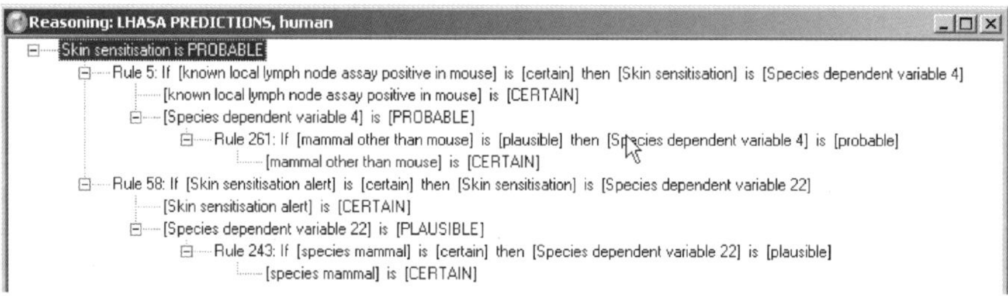

Figure 13.4   Reasoning for the prediction of the human skin sensitisation potential of benzoyl chloride.

chloride has been tested in the assay and gave a positive result (if you were running the program and asked to view example 2, listed in Figure 13.3, you would see the data) and human belongs to the set [mammal other than mouse]. So rules 5 and 261 taken together lead to the conclusion that skin sensitisation in humans is probable. Rules 58 and 243 apply to this query just as they did to the previous one, since this query is also a carboxylic acid chloride, and they predict skin sensitisation to be plausible in humans. Applying the rule that says that the strength of a proposition is equal to the strongest force of the arguments supporting it, the program concludes that skin sensitisation in humans is

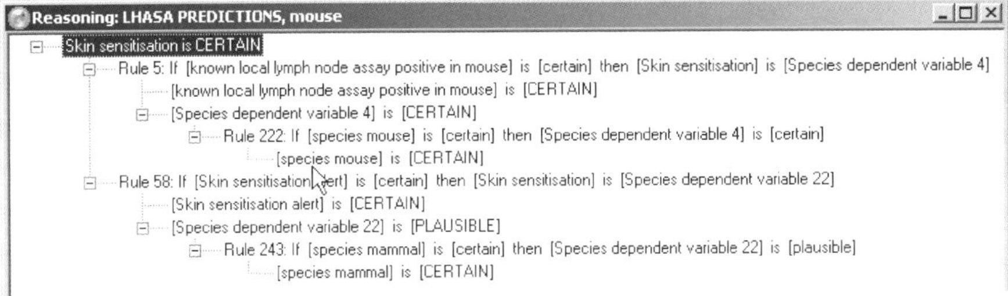

Figure 13.5 Reasoning for the prediction of the mouse skin sensitisation potential of benzoyl chloride.

probable, that being higher ranked than plausible. Figure 13.5 shows the reasoning for the mouse. [Species mouse] is, of course, certain and if you work through the reasoning you will find that the output from rule 5 becomes certain. Certain is stronger than plausible and so skin sensitisation in mouse is reported to be certain. Rules 222 and 261 will both have been considered in each of the analyses but, as mentioned earlier, only rules that contributed to the prediction of interest are displayed. Consider, for example, what happens with rule 261 if you have asked for a prediction for human skin sensitisation: the value for [species mouse] is impossible, and so the output from rule 261 is open.

There are rules about properties such as octanol–water partition coefficient and skin permeability in Derek for Windows, but they do not appear in the reasoning reports used in the examples in this chapter, because they did not modify the predictions for the chosen structures and end point.

13.5 The Derek for Windows Alert Editor

Alerts in Derek for Windows are not described by means of linear codes. They are represented graphically.[18] Derek for Windows alert number 401, for skin sensitisation associated with carboxylic acid chlorides, is shown in Figure 13.6. The box on the left contains the description that is displayed to an end-user of the substructural features that trigger the alert. The comments field allows the knowledge base writer to communicate supporting information to the end-user – about the toxicological mechanism, for example. If you clicked the tab labelled "Validation" you would see information about how the alert performed against various test data sets. User organisations can add their own comments, under the tab "Custom Comments". Comments under "Rule Writer Comments" can be viewed in editor mode but are not displayed to end-users. In this case, four literature references are available describing studies on which the alert is based, and four specific examples of chemicals containing the alert, with the relevant toxicological data, are available for viewing.

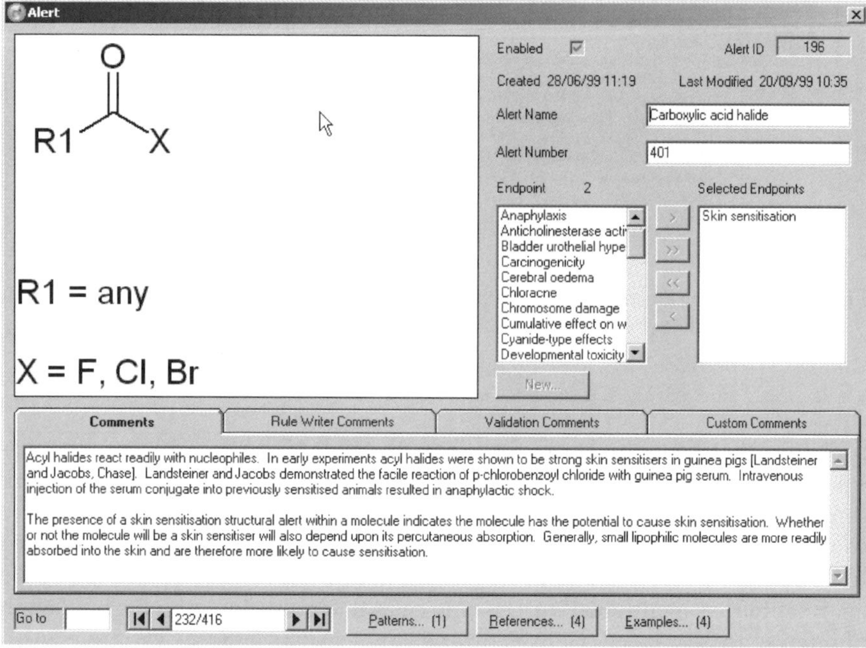

Figure 13.6   Alert 401 in Derek for Windows for skin sensitisation associated with carboxylic acid chlorides.

The button labelled "Patterns . . . " allows a knowledge base writer to view and edit the descriptions of alerts that drive the program, the diagram to the top left of the window being only an image for display to end-users, but as you will guess from the diagram, the alert for carboxylic acid chlorides is not very interesting. Figure 13.7 shows the much more complicated alert for the mutagenicity of aromatic amines and amides. You can use multiple patterns to describe an alert. In this case there are nineteen, one of which is the one displayed in Figure 13.8. The upper box shows the pattern. For some of the atoms in the diagram, the knowledge base writer specifies that they must satisfy certain criteria. If you select one of these atoms, the tabbed boxes on the right show its attributes. The pattern also contains some generic groups, R1 and R2. I have selected group R1 and so its definition appears at the bottom of the screen.

The first sub-pattern shows that R1 is a carbonyl group to which anything else can be attached. The star beside the carbon atom means that this is the point of attachment to the main substructure. I have not selected that carbon in the picture I chose for this figure, but were you to do so you would find that it is allowed to be connected to only two heteroatoms. So the pattern is one for an amide. R1 is further qualified by the second sub-pattern, for which the "exclusion" box is checked. I have selected the carbon atom in this sub-pattern and its attributes are shown on the right. Fragments in which less than two

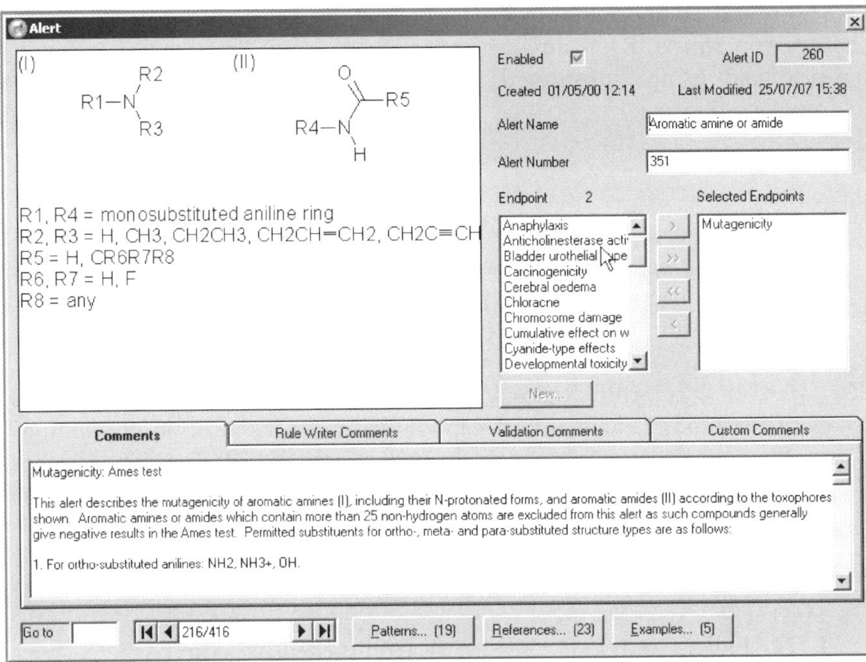

Figure 13.7 Alert 351 in Derek for Windows for mutagenicity associated with aromatic amines and amides.

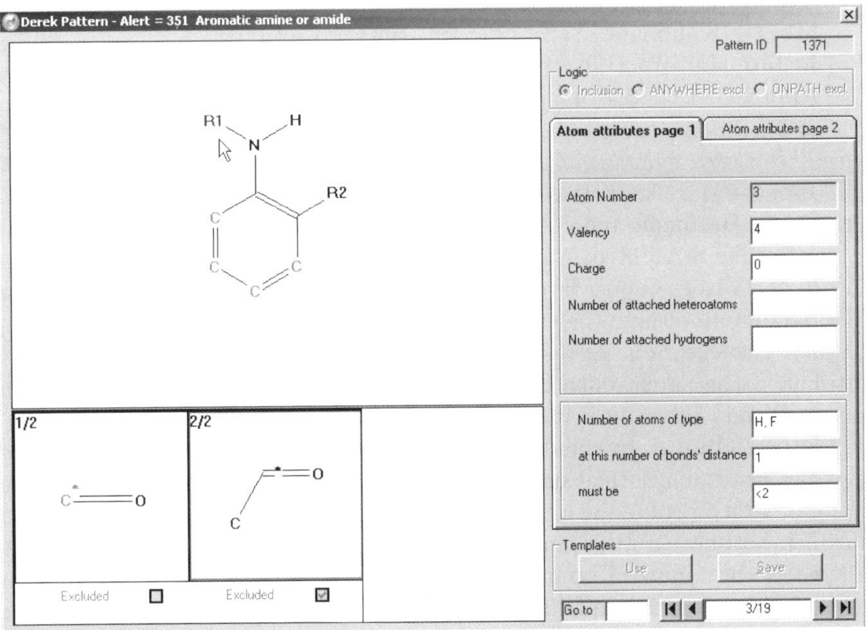

Figure 13.8 One of the patterns for alert 351 in Derek for Windows.

hydrogen and/or fluorine atoms are attached to the atom are excluded. So, expressed in words, R1 is an acyl group which is not branched at the alpha position and is not substituted at that position by anything other than hydrogen or fluorine.

References

1. M. P. Smithing and F. Darvas, HazardExpert – an Expert System for Predicting Chemical Toxicity, in *Food Safety Assessment*, ed. J. W. Finley, S. F. Robinson, D. J. Armstrong, ACS Symposium Series Vol. 484, Am. Chem. Soc., Washington DC, 1992, pp. 191–200.
2. *"More or Less"*, BBC Radio 4, broadcast at 13.30 on 26th December 2008.
3. R. Mannhold, Calculation of Lipophilicity: a Classification of Methods, in *Pharmacokinetic Profiling in Drug Research*, ed. B. Testa, S. Krämer, H. Wunderli-Allensprach and G. Folkers, Wiley–VCH, Weinheim, 2006, pp. 333–352.
4. P. J. Krause, S. Ambler, M. Elvang-Gøransson and J. Fox, A Logic of Argumentation for Reasoning Under Uncertainty, *Comput. Intell.*, 1995, **11**(1), 113–131.
5. P. N. Judson and J. D. Vessey, A Comprehensive Approach to Argumentation, *J. Chem. Inf. Comput. Sci.*, 2003, **43**, 1356–1363.
6. M. Elvang-Gøransson, P. J. Krause and J. Fox, Dialectic Reasoning with Inconsistent Information, in *Uncertainty in Artificial Intelligence: Proceedings of the 9th Conference*, ed. D. Heckerman and A. Mamdani, Morgan Kaufmann, San Francisco, 1993, pp. 114–121.
7. J. Fox, D. W. Glasspool and J. Bury, Quantitative and Qualitative Approaches to Reasoning under Uncertainty in Medical Decision Making, in *8th Conference on Artificial Intelligence in Medicine in Europe, AIME 2001 Cascais, Portugal, July 2001, Proceedings*, ed. S. Quaglini, P. Barahone and S. Andreassen, Springer, Berlin, 2001, pp. 272–282.
8. D. K. Hardman and P. Ayton, Arguments for Qualitative Risk Assessment: the StAR Risk Adviser, *Expert Systems*, 1997, **14**(1), 24–36.
9. P. N. Judson, C. A. Marchant and J. D. Vessey, Using Argumentation for Absolute Reasoning About the Potential Toxicity of Chemicals, *J. Chem. Inf. Comput. Sci.*, 2003, **43**, 1364–1370.
10. P. J. Krause, J. Fox and P. N. Judson, An Argumentation-Based Approach to Risk Assessment, *IMA J. Math. Appl. Bus. Ind.*, 1993–4, **5**, 249–263.
11. P. N. Judson, J. Fox and P. J. Krause, Using New Reasoning Technology in Chemical Information Systems, *J. Chem. Inf. Comput. Sci.*, 1996, **36**, 621–624.
12. J. J. Langowski, P. N. Judson, M. Patel and C. A. G. Tonnelier, StAR. A Knowledge-Based Computer System for Carcinogenic Risk Assessment, in *Animal Alternatives, Welfare and Ethics. Developments in Animal and Veterinary Sciences, Vol. 27*, ed. L. F. M. Van Zutphen and M. Balls, Elsevier Science, 1997, pp. 747–752.

13. M. D. Barratt, D. A. Basketter, M. Chamberlain and G. D. Admans, An Expert System Rulebase for Identifying Contact Allergens, *Toxicol. In Vitro*, 1994, **8**, 1053–1060.
14. M. D. Barratt, The Role of Structure-Activity Relationships and Expert Systems in Alternative Strategies for the Determination of Skin Sensitisation, Skin Corrosivity, and Eye Irritation, *Alternat. Lab. Animals*, 1995, **23**, 111–122.
15. M. D. Barratt and J. J. Langowski, Validation and Subsequent Development of the DEREK Skin Sensitisation Rulebase by Analysis of the BgVV List of Contact Allergens, *J. Chem. Inf. Comput. Sci.*, 1999, **39**, 294–298.
16. K. Langton and C. A. Marchant, Improvements to the Derek for Windows Prediction of Chromosome Damage, *Toxicol. Lett.*, 2005, **158**, S36–S37.
17. Derek for Windows is developed by, and available from, Lhasa Limited, 22-23 Blenheim Terrace,Woodhouse Lane, Leeds LS2 9HD, United Kingdom.
18. C. A. G. Tonnelier, J. Fox, P. N. Judson, P. J. Krause, N. Pappas and M. Patel, Representation of Chemical Structures in Knowledge-Based Systems: the StAR System, *J. Chem. Inf. Comput. Sci.*, 1997, **37**, 117–123.

CHAPTER 14

Predicting Metabolism

Very often the toxic effects of a chemical are actually due, not to the chemical itself, but to one or more of its metabolites. In some cases the metabolites have been detected and studied. In others the evidence is there, but it is not known what the metabolites are. There are many metabolic pathways in living cells that interconvert chemicals that are normal constituents of cells – endobiotic chemicals. Chemicals foreign to living systems – xenobiotic chemicals – may be, so to speak, mistaken for endobiotic chemicals and metabolised, but specific metabolism of xenobiotic chemicals is more important. Xenobiotic chemicals pose a constant threat to living systems. Many types of cells have enzymes and mechanisms for disposing of xenobiotic chemicals, but in mammals liver cells in particular are designed to deal with the stream of potentially damaging chemicals absorbed into the body through the gut.

The Ames test, devised by Bruce Ames and colleagues,[1,2] is an example of a procedure in which activation through metabolism can be observed, but which gives no information about the metabolites. The purpose of the test is to detect the potential mutagenicity of chemicals. Most often, strains of *Salmonella typhimurium* bacteria are used which, because of a genetic defect, are unable to synthesise histidine, and therefore depend on being supplied with it for growth. The bacteria are put into a medium containing the chemical and given no histidine. If the chemical causes mutations in the bacteria, mutants that regain the ability to synthesise histidine will multiply. So the appearance of successful, growing colonies of the bacteria indicates that the chemical is mutagenic, at least to *Salmonella typhimurium*. The test is usually done both in the absence and the presence of a rat liver extract, termed S9, which contains many of the enzymes responsible for the metabolic breakdown of chemicals in the liver.[1,3] It is frequently found that a chemical shows little or no mutagenic effect in the absence of S9 but is mutagenic in the presence of S9.

It is not necessary to be explicit about metabolism in order to predict toxicity on the basis of structural alerts. Derek for Windows, M-Case, and other

RSC Theoretical and Computational Chemistry Series No. 1
Knowledge-based Expert Systems in Chemistry: Not Counting on Computers
By Philip Judson
© Philip Judson 2009
Published by the Royal Society of Chemistry, www.rsc.org

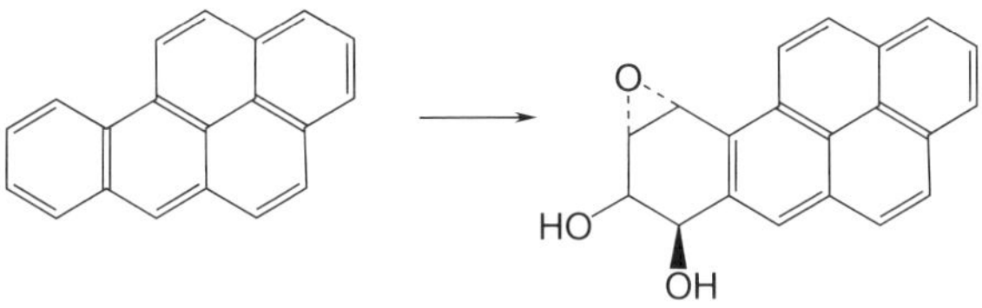

Figure 14.1 Likely metabolic oxidation of benzo[*a*]pyrene.

systems predict the carcinogenicity of polyaromatic hydrocarbons, such as benzo[*a*]pyrene (see Structure 14.1 in Figure 14.1). Benzo[*a*]pyrene, found in coal tar and cigarette smoke, is classed as a carcinogen. Its mode of action involves metabolic conversion, most probably to the diol epoxide (see Structure 14.2) which can become intercalated into DNA (*i.e.* inserted between the "threads" of the helix like a coin in a slot), and bind covalently through reaction of the epoxide group with an amine group in the DNA. This disrupts DNA replication, resulting in mutations some of which lead to the creation of cancerous cells. The shape of a polycyclic hydrocarbon determines whether it shows this kind of toxicity. It must be planar for intercalation into DNA to be possible, and it must fit the P-450 enzymes responsible for its oxidation in such a way that an epoxide group is formed, survives, and is in the right position to be close to an amine when the metabolite is trapped in the DNA. The alert for this kind of toxicity in Derek for Windows, for example, describes the kinds of polyaromatic hydrocarbon that observation has shown to have the right shapes and sizes to give rise to active metabolites, and distinguishes them from those that are not expected to do so.

Derek for Windows "knows" what a potentially carcinogenic hydrocarbon looks like, but it does not "know" that the actual toxin is a metabolite. Probably one third or more of the alerts in Derek for Windows describe substructures that are believed to undergo metabolic transformation into reactive groups that are actually responsible for toxicity. It would be intellectually more satisfying if Derek for Windows simply contained alerts truly associated with toxicity, and a different module predicted their metabolic introduction into structures that were not necessarily toxic in themselves. It might lead to improved predictions, since models normally perform better, the closer they are to what they model. It would also make knowledge base development and maintenance easier, because knowledge base writers would not need to think through the metabolic possibilities and their implications. This chapter is about computer systems for predicting the metabolism of xenobiotic chemicals.

14.1 COMPACT, MetaSite and SPORCalc

Three systems for predicting mammalian metabolism which are not knowledge based as defined in this book, but which have things in common with the knowledge-based approach have been described by David Lewis,[4] Gabriele Cruciani and co-workers,[5] and Scott Boyer and co-workers.[6]

A group of enzymes known collectively as CYP-450 (for "cytochrome P 450") is important in drug research. Being the enzymes primarily responsible for metabolic oxidations of compounds in the liver, they are responsible for the degradation of pharmaceutically active compounds, for the activation of some pro-drugs (compounds that are not active in themselves but give rise to active metabolites), and for the unwitting creation of toxic metabolites. The safe, effective dose of a drug may be higher than it would otherwise be, because it is partly converted by one of the CYP-450 isozymes into an inactive metabolite. If a patient is given this drug together with another one that competes for the same metabolic site, greater amounts of each drug may survive leading to harmful side effects – so-called drug–drug interaction. So being able to model the CYP-450 oxidation of drugs is an important goal.

COMPACT[4] is a methodology, not a discrete piece of software. The structures of an increasing number of CYP-450 enzymes have been determined. The approach is to model the binding of a query structure to the active sites of the enzymes, using 3D molecular modelling methods, in order to get guidance on which isozymes are most likely to be responsible for oxidation of the query, and thus what the products might be and how likely it is that drug–drug interaction problems will be seen.

MetaSite[5,7] is, in effect, a software implementation of this approach, but with a slightly difference emphasis. Although it gives guidance on which isozyme is most likely to interact with a query molecule, the designers of the program recognise that this is a difficult thing to do. They concentrate more on trying to predict the likely site of oxidation in the query structure when the isozyme responsible is already known or suspected.

Boyer *et al.*[6] describe how they use data mining of the MDL Metabolite database[8] to generate fingerprints which can subsequently be used to predict the more likely sites of metabolic attack on chemicals, using what they call a Substance Product Occurrence Ratio Calculator (SPORCalc). Fingerprints (see Chapter 6.3) were derived for all atoms in all substrates and for all reacting centres in the database. When a query structure is entered both sets are searched for fingerprint matches to every site in the query. From the ratio of occurrence of the sites in the two sets from the database – those where reaction took place and those where it did not – the frequencies of reactions occurring at each site can be ranked. It is thus possible to list the sites in the query from the most to least likely sites for metabolism. Given that information, a chemist can consider ways to modify the structure of a compound under consideration as a pharmaceutical to change its susceptibility to P-450 metabolism.

14.2 XENO, MetabolExpert and META

Predicting the metabolism of xenobiotic chemicals with a knowledge-based system was pioneered by Todd Wipke[9] and colleagues, who developed XENO – a program based on the SECS system (see Chapter 3.1.1), which had its origins in the LHASA project at Harvard. Ferenc Darvas[10] and, a little later, Gilles Klopman and colleagues[11,12] were also early researchers in the field. Wipke's XENO, Darvas' MetabolExpert and Klopman's META all use a knowledge base of reactions, or "biotransformations". Each biotransformation description includes the substructural fragment that must be present in a structure for it to undergo the reaction, and the information needed for the computer program to generate the structure of the appropriate product from the query structure.

The knowledge base for META contains two kinds of biotransformations – those promoted by enzymes and spontaneous chemical reactions (usually occurring because an enzymatic conversion has created an unstable product). Figure 14.2 shows an example used by Klopman *et al.* in their first paper about META, in which the overall result is the dealkylation of methylethyl-*N*-nitrosamine (see Structure 14.3).[11] The presence of a -CH_2-N- group in the starting material triggers a biotransformation for the hydroxylation of an amine by a P-450 enzyme to generate a product containing a –CH(OH)–N– group. The resultant product in this case contains the trigger for a spontaneous reaction in the META dictionary describing a tautomeric shift leading to decomposition into two fragments. The nitrogen containing fragment is a tautomer of methyl-*N*-nitrosamine, and so the overall effect is one of amine dealkylation, creating Structure 14.4.

The inclusion of spontaneous chemical changes in META is important, and the lack of them, or limited coverage, in some other systems has been a weakness. For example, early versions of Meteor (see Chapter 14.3) contained only biotransformation reactions, on the grounds that biotransformation was what it was supposed to be about. I was in the audience at a presentation when a non-chemist demonstrating to non-chemists entered benzoyl chloride for processing. The program happily generated ring hydroxylation products such as 4-hydroxybenzoyl chloride and then derivatives of those, all still containing the acyl chloride group. Members of the audience who did not notice anything amiss were being misled: members of the audience who did notice, presumably wondered why the program failed to report the obvious, but were too polite to ask. What was needed in the knowledge base, was the chemical transformation

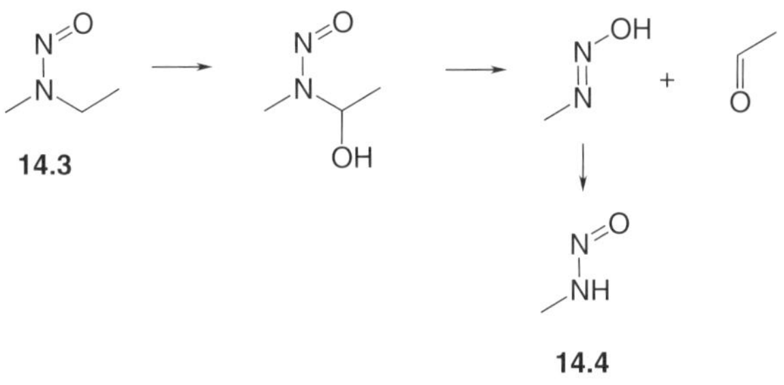

Figure 14.2 Dealkylation of a nitrosamine.

of acyl halides into carboxylic acids with a high likelihood assigned to it so that it would take priority, and it is now included in Meteor, leading to the creation of benzoic acid as the first level product.

XENO was not developed much beyond the original prototype, but MetabolExpert and META include some controls over the growth of the metabolic tree. In MetabolExpert, probabilities are attached to biotransformation descriptions like they are in HazardExpert (see Chapter 10.3) and these are used to direct the development of the metabolic tree. How favoured a biotransformation is, can also take account of physicochemical properties of the substrate, such as its estimated log P. In META, each biotransformation is assigned a priority based on how prevalent it is considered to be by metabolism experts. Recognising that assigning priorities is a difficult task which gets harder the bigger the dictionary becomes, Klopman's group have experimented with using genetic algorithms to automate the assignments.[13]

Both MetabolExpert[14] and META[15] are commercially available at the time of writing of this book. So is CATABOL, a system originally developed for predicting the environmental degradation of chemicals and now extended to predict mammalian metabolism. CATABOL is described in Chapter 16.2.

14.3   Meteor

Even by the mid 1990s there was still scepticism in the broader scientific community about the feasibility of predicting metabolism. The general view was that it was relatively easy to suggest all the possible metabolites of a compound, but that it would be impossible to restrict the predicted metabolic tree to the metabolites likely to be detected in practice.

Lhasa Limited was still based in the chemistry department at Leeds University, sharing offices with the LHASA research group there. Seeing at first hand that the LHASA program could control the combinatorial explosion associated with synthesis planning – a problem at least as difficult as the one of controlling the size of a metabolic tree – we believed something could be achieved. Linking metabolism and toxicity prediction offered a first step: instead of asking the question "What metabolites of the user's query compound are most likely to be observed?" we could ask "Is it possible to generate alerts associated with the toxicological end-point of interest from the user's query compound?" Together with the LHASA research group in Leeds we sought funding from the Science and Engineering Research Council, but the committee considering the application turned it down on the grounds that creating a metabolism prediction system was not a realistic goal.

The development of what became Derek for Windows through the StAR project (see Chapter 13), was the first step in a plan for staged development of reasoning-based systems at Lhasa Limited. Predicting toxicity requires the processing of knowledge about chemical structures. Predicting metabolism requires the processing of knowledge about sets of inter-related structures – starting materials and products. The step beyond that is to reason about metabolic trees – sets of reactions – and perhaps even sets of metabolic trees. The beginnings of reasoning about metabolic trees feature a little in Chapter 18, but it is an area not yet much unexplored. Once Derek for Windows had been created, it was time to try again for sponsorship of a metabolism system, and this time it was successful. Half a dozen companies agreed to sponsor a three year project, leading to the creation of "Meteor".[16,17] The steering committee for the project was made up of staff from the metabolism research departments of the sponsoring companies, and their interests went a lot broader than just supporting toxicology work. It quickly became apparent that we needed to develop a standalone application for metabolism, not just one to support toxicity prediction.

Figure 14.3 shows one of the biotransformation descriptions in Meteor version 10.0.2 for hydrolysis of amides, biotransformation number 152, as displayed to an end user and in the knowledge base editor. As with alert descriptions in Derek for Windows, the diagram in the biotransformation description is just an image for display to the user. The descriptions that drive the program are contained in "patterns", one of which is shown in Figure 14.4. The knowledge base writer who enters a pattern, specifies the features needed in a reactant to trigger the biotransformation. Attributes can be attached to atoms and bonds as they can be in Derek for Windows, for example, to specify that an atom or bond must or must not be aromatic, or in a ring or chain, or that neighbouring atoms must, or must not, be certain elements, and the writer of

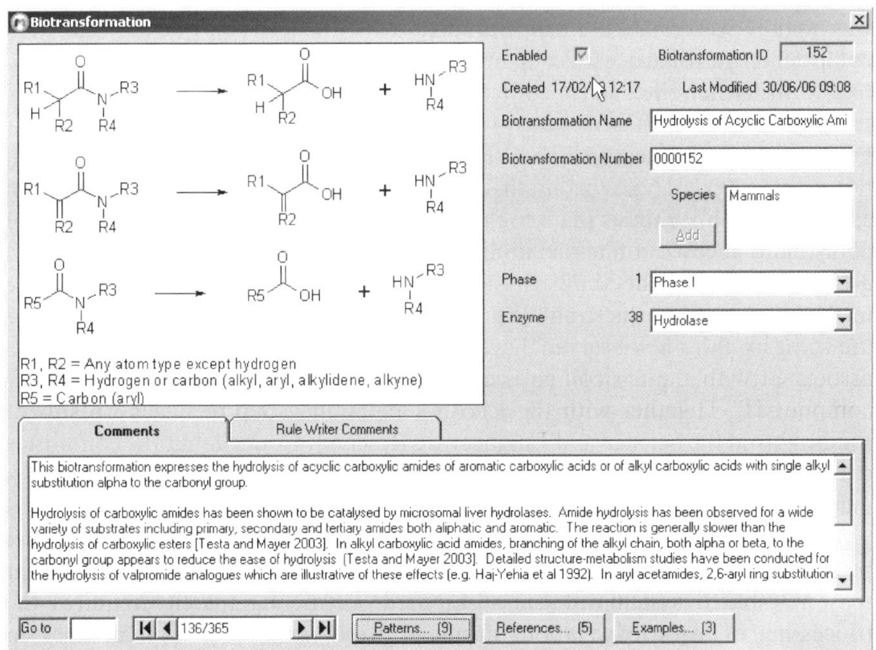

Figure 14.3   Meteor biotransformation number 152.

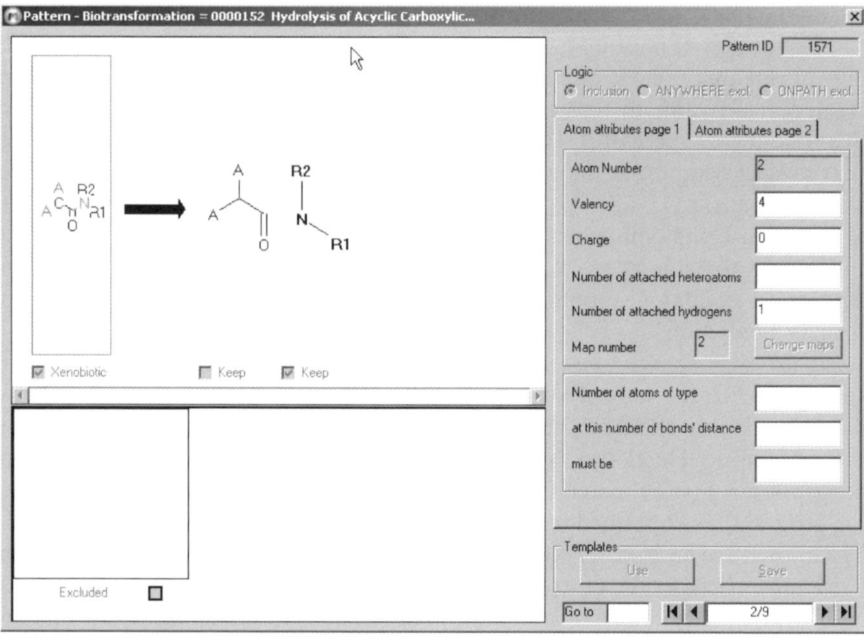

Figure 14.4   A biotransformation pattern in Meteor.

the biotransformation maps each atom in the starting substructure to the corresponding atom in the product(s).

Given ethyl benzoate as a query, Meteor generates the display shown in Figure 14.5 when options for controlling the size of the metabolic tree are set to their default values. The display of the metabolic tree can be expanded and the first part of it is shown in Figure 14.6 (someone using the program can scroll the display up and down on the screen to see the other parts of it). Biotransformation number 152 appears on the right and is considered to be a plausible one for the query compound. Terminal hydroxylation of the ethyl group (biotransformation 73, on the left) is also considered to be plausible. The terms for expressing the likelihood that a biotransformation will take place are the same as the ones used for likelihood of toxicity in Derek for Windows – certain, probable, plausible, equivocal, doubted, improbable and impossible – and one of the default options is that only metabolites that are at least plausible are displayed.

Another default option is that the products of phase 2 metabolic transformations are not further metabolised. One of the ways in which the body disposes of xenobiotic chemicals is to render them water soluble so that they can be excreted, for example by attaching a sugar residue to the chemical. For that to be possible, it is often necessary first to introduce a suitable functional group (this is not to imply that the liver thinks to itself "aha, better set something up here so that I can attach a sugar" – what it means is that evolution has favoured doing the right kinds of reactions). Figure 14.7 shows a reaction sequence from

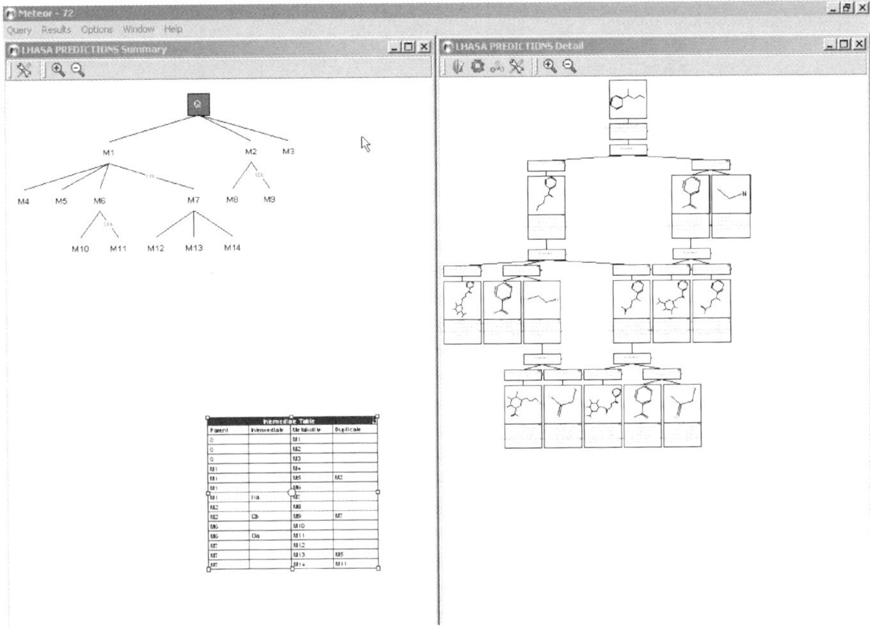

Figure 14.5 A Meteor metabolic tree for ethyl benzoate.

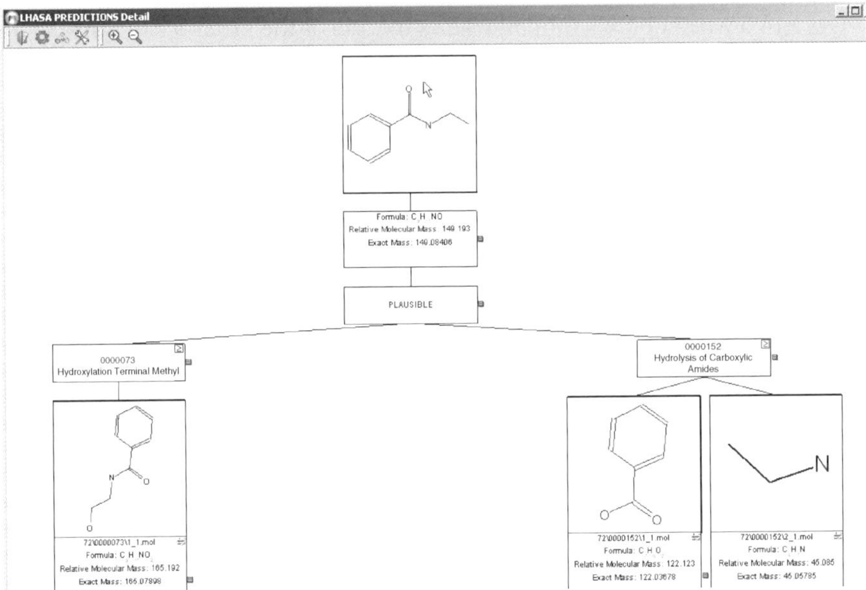

Figure 14.6 Part of the tree from Figure 14.5 in more detail.

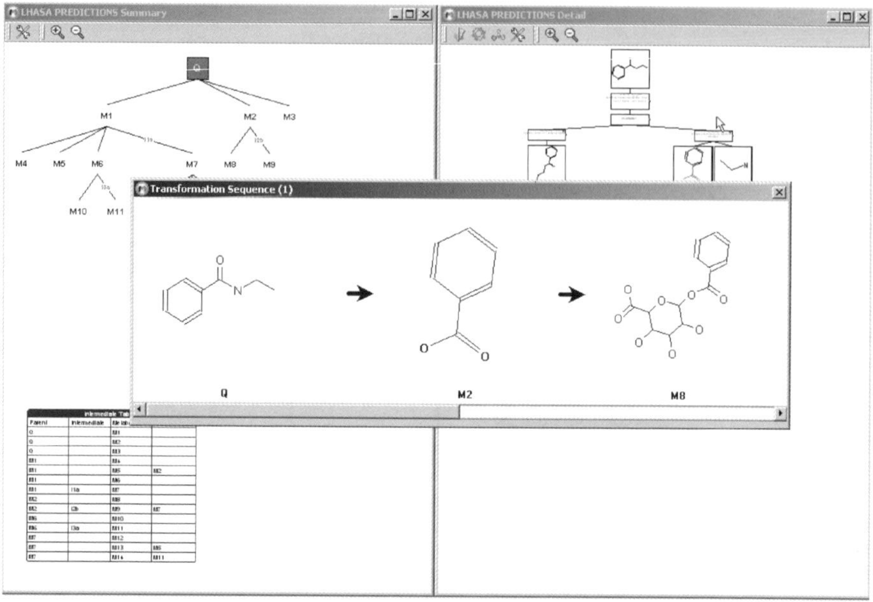

Figure 14.7 A reaction sequence leading to a phase 2 metabolite.

the metabolic tree generated by Meteor for ethyl benzoate. The first step, a phase 1 metabolic reaction, is the creation of a carboxylic acid. The second step, a phase 2 metabolic reaction, is the formation of a glucuronide. The product contains three alcoholic hydroxyl groups and one new carboxylic acid group, all of which could be glucuronidated again, and the same would apply to all those products. Blocking the growth of branches of the metabolic tree beyond phase 2 metabolites prevents this. In laboratory studies, a second glucuronidation, and even a third, is occasionally observed, but it is usually a rare or minor event. The increased hydrophilicity of a glucuronidated product means both that it binds poorly to the enzymes that promote metabolism, and that it is easily excreted and thus removed too quickly to be metabolised further.

Meteor is twinned with Derek for Windows in a single software package, but the applications are designed and developed to be able to work on their own. A user can request that a structure in Derek for Windows be transferred to Meteor for processing, or that a metabolite in Meteor be passed to Derek for Windows for a toxicity prediction, but the idea of restricting Derek for Windows to directly-acting alerts and letting Meteor generate and return metabolites automatically to cover metabolic activation, has not been implemented. The reasons are practical ones. Generating metabolites would increase processing time, making Derek for Windows less convenient to use, and it would be for limited benefit. It is true that metabolic routes would sometimes be discovered to toxic structures, that neither the user nor Derek for Windows knowledge base writers had thought of, but probably not very often. The important routes are already covered implicitly in Derek for Windows by

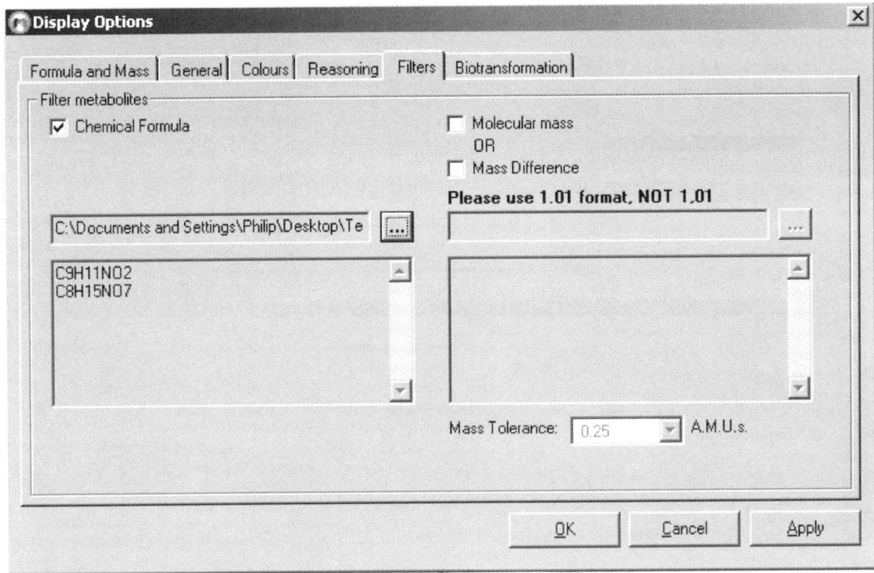

Figure 14.8 Importing chemical formulae from a file into Meteor.

alerts and rules that define the precursors to toxins, like the polyaromatic hydrocarbons mentioned earlier in this chapter, and users can check for unexpected toxic metabolites by running Meteor themselves and transferring the results to Derek for Windows if they wish. Whether to automate the prediction of metabolic activation remains an option which is regularly reviewed. The time to make the change will be when the cost in reduced speed is outweighed by the benefits of a more powerful prediction model, and that time may be getting close.

In practice, Meteor is used to support metabolism studies more than to support toxicity prediction. Analysis using gas chromatography and mass spectrometry (GCMS) often throws up peaks for masses that are hard to interpret. Is a peak associated with a metabolite or not? If it is, what is its structure? Answering this question, presents a way of limiting the metabolic tree generated by Meteor very effectively without losing relevant predictions, however unlikely they may seem. A feature in Meteor allows the user to import data from a file listing the molecular formulae or exact or relative masses of structures of interest (see Figure 14.8). Meteor generates all possible metabolic paths (subject to any constraints that the user may have set), but retains only those that lead to structures with formulae or masses corresponding to the ones of interest to the user. In this example, the tree is reduced to the single sequence shown in Figure 14.9. Although the information is not captured in this screen shot, the formulae to which the search was directed, $C_9H_{11}NO_2$ and $C_8H_{15}NO_7$, are those of Metabolites M1 and M10 in Figure 14.9. Metabolite M6 is also displayed, of course, because it is on the route to M10 and thus part of the explanation of how M10 may have been formed.

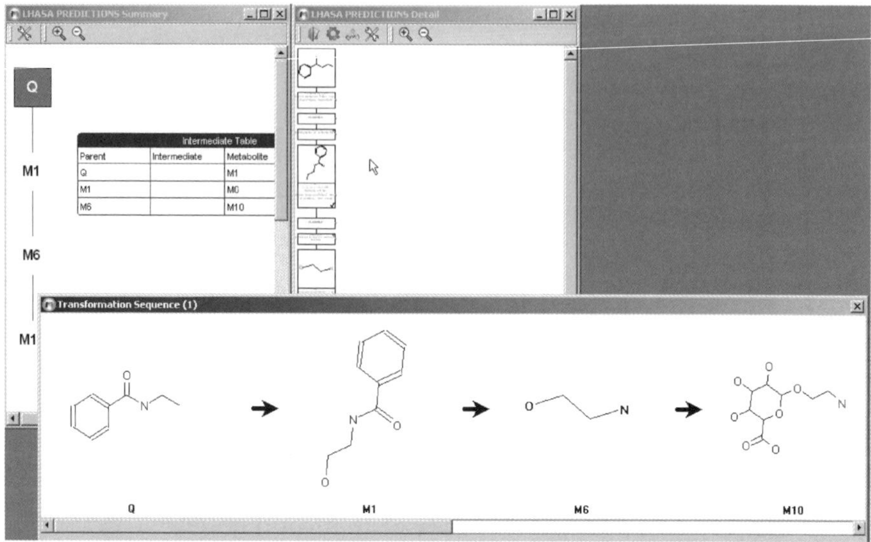

Figure 14.9 A metabolic sequence consistent with the formulae in Figure 14.8.

There are difficulties with assessing just how well a metabolism prediction system performs. Users want to know what they are likely to observe, but the compounds you detect experimentally, are rarely all of those that have been formed. However, while predictions that Meteor makes are a long way short of 100% correct, they are now close enough to what is observed in practice to be useful. Like the teams working on MetabolExpert and META, knowledge base development staff assess the likelihood that each biotransformation will be seen in practice, a task in which they were assisted during the original development of Meteor by the members of the project steering committee, all of whom had a good deal of practical knowledge about metabolism. We are fortunate also to have a lot of help from two experts in mammalian metabolism, David Hawkins[18] and Bernard Testa.[19,20]

Early in the development of Meteor, it became evident that we had missed something important in our thinking about the use of reasoning. There is more to how Meteor chooses which biotransformations to display than this chapter describes. If you look again at Figure 14.6 you will notice a small " $\geq$ " symbol at the top right of the box for biotransformation 73. It reveals that biotransformation 73 was selected for display in preference to some others. Why and how, are the topics of the next chapter.

References

1. K. Mortelmans and E. Zeiger, The Ames Salmonella/Microsome Mutagenicity Assay, *Mutat. Res.*, 2000, **455**, 29–60.

2. B. N. Ames, F. D. Lee and W. E. Durston, An Improved Bacterial Test System for the Detection and Classification of Mutagens and Carcinogens. *Proc. Natl. Acad. Sci.*, 1973, **70**, 782–786.
3. J. McCann, N. E. Spingarn, J. Kobori and B. N. Ames, Detection of Carcinogens as Mutagens: Bacterial Tester Strains with R Factor Plasmids. *Proc. Natl. Acad. Sci.*, 1975, **72**, 979–83.
4. D. F. V. Lewis, COMPACT: a Structural Approach to the Modelling of cytochromes P450 and Their Interactions with Xenobiotics, *J. Chem. Tech. Biotech.*, 2001, **76**, 237–244.
5. G. Cruciani, E. Carosati, B. De Boeck, K. Ethirajulu, C. Mackie, T. Howe and R. Vianello, Metasite: Understanding Metabolism in Human Cytochromes from the Perspective of the Chemist, *J. Med. Chem.*, 2005, **48**, 6970–6979.
6. S. Boyer, C. H. Arnby, L. Carlson, J. Smith, V. Stein and R. C. Glen, Reaction Site Mapping of Xenobiotic Biotransformations, *J. Chem. Inf. Model*, 2007, **47**, 583–590.
7. G. Caron, G. Ermondi and B. Testa, Predicting the Oxidative Metabolism of Statins: an Application of the MetaSite Algorithm, *Pharm. Res.*, 2007, **24**, 480–501.
8. MDL Metabolite comes from Symyx Technologies, Inc., 3100 Central Expressway, Santa Clara, CA 95051, USA.
9. W. T. Wipke, G. I. Ouchi and J. T. Chou, Computer, -Assisted Prediction of Metabolism, in *Structure–Activity Correlation as a Predictive Tool in Toxicology: Fundamentals, Methods, and Applications*, ed. Leon Golberg, Hemisphere, Washington, DC, 1983, pp. 151–169.
10. F. Darvas, METABOLEXPERT an Expert System for Predicting Metabolism of Substances, in *QSAR. Environmental Toxicology, Proceedings of an International Workshop 1986*, ed. K. L. E. Kaisler, Reidel, Dordrecht, 1987, pp. 71–81.
11. G. Klopman, M. Dimayagu and J. Talafous, META. 1. A Program for the Evaluation of Metabolic Transformations of Chemicals, *J. Chem. Inf. Comput. Sci.*, 1994, **34**, 1320–1325.
12. J. Talafous, L. M. Sayre, J. J. Mieyal and G. Klopman, META. 2. A Dictionary Model of Mammalian Xenobiotic Metabolism, *J. Chem. Inf. Comput. Sci.*, 1994, **34**, 1326–1333.
13. G. Klopman, M. Tu and J. Talafous, META. 3. A Genetic Algorithm for Metabolic Transform Priorities Optimisation, *J. Chem. Inf. Comput. Sci.*, 1997, **37**, 329–334.
14. MetabolExpert comes from CompuDrug International Inc., 115 Morgan Drive, Sedona, AZ 86351, USA.
15. META comes from Multicase Inc, 23811 Chagrin Blvd Ste 305, Beachwood, OH, 44122, USA.
16. N. Greene, P. N. Judson, J. J. Langowski and C. A. Marchant, *SAR QSAR Environ. Res.*, 1999, **10**, 299–314.
17. B. Testa, A.-L. Balmat, A. Long and P. Judson, Predicting Drug Metabolism – an Evaluation of the Expert System METEOR, *Chem. Biodiversity*, 2005, **2**, 872–885.

18. D. R. Hawkins (ed.), *Biotransformations*, an annual series from the Royal Society of Chemistry, Cambridge, England, starting in 1989.
19. B. Testa and S. D. Krämer, *The Biochemistry of Drug Metabolism*, Wiley–VCH, Weinheim, 2008.
20. B. Testa and P. Jenner, *Drug Metabolism: Chemical and Biochemical Aspects*, Marcel Dekker, New York, 1976.

CHAPTER 15

Relative Reasoning

When asked to classify biotransformations according to the likelihood that they would be seen in practice, metabolism experts were able to do that to some extent for many biotransformations, but one response soon became familiar to us. "You are asking us the wrong question. It would be easier to answer the question: 'Is this biotransformation more likely than that one?'"

What happens in practice is largely determined by competition between different potential biotransformations, and much of what is seen in metabolism depends upon reaction kinetics. If your starting material is susceptible to metabolism at more than one site, which product predominates will depend on the rates of the reactions at the different sites. The current state of knowledge about how to calculate reaction rates, especially for reactions promoted by enzymes, does not allow the calculation of even half-trustworthy rates for novel structures in most cases. But for practical purposes it is not necessary to do that if you can come up with rules of thumb of the kind "when biotransformations X and Y compete, X usually predominates". It was this kind of reasoning that metabolism experts were able to do, and which needed to be brought into Meteor.

The likelihood that a biotransformation will be seen even in the absence of competition does vary, of course. For example, how well a compound fits the site of the enzyme that promotes a biotransformation will influence the likelihood of seeing the biotransformation in practice, quite apart from the influences of competition with other enzymes. Let us use the terms "absolute reasoning" and "relative reasoning" to refer to rules of the kind "A is probable" and "A is more likely than B", respectively. Is absolute or relative reasoning more suitable for the prediction of metabolism, or are both needed?

Imagine a domain in which only ten events are possible, each is differently favoured, and the exact absolute likelihood of each of them is known (see Figure 15.1). From Figure 15.1, based on absolute likelihood, you can find out also about relative likelihood – that 'b' is more likely than 'c', and that 'c' is more

RSC Theoretical and Computational Chemistry Series No. 1
Knowledge-based Expert Systems in Chemistry: Not Counting on Computers
By Philip Judson
© Philip Judson 2009
Published by the Royal Society of Chemistry, www.rsc.org

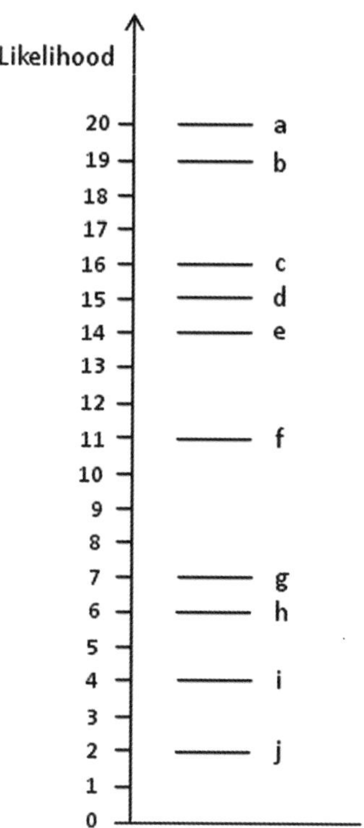

Figure 15.1   A domain in which the likelihood of every event is known.

likely than 'h', for example. More precisely, in terms of the units used in Figure 15.1, 'b' is 3 units more likely than 'c' and 'c' is 10 units more likely than 'h'. Now imagine that you are not given the absolute levels of likelihood, but instead the relationships between pairs of events – 'a' is more likely than 'b', 'b' is more likely than 'c', and so on. From this knowledge about relative likelihood you can construct Figure 15.2. Suppose that you know precisely how much more likely each thing is than the one below it in the stack, as indicated in Figure 15.2. If the units of measurement are the same as the ones used in Figure 15.1, then the pictures of the world presented in the two figures are identical. They differ only in how the diagrams are labelled.

The real world is not like that. We do not know the exact likelihood of every event. We only have approximations and many events may be ranked as equally likely within the limits of those approximations. Neither do we know precisely how much more likely one event is than another. At best we may know that one is a bit more likely or a lot more likely than the other. We know nothing at all about the relationships between some events. Finally, we do not know about all events in the world of metabolism. If we did, we would not be in need of a prediction system. In place of Figure 15.1, we now have perhaps

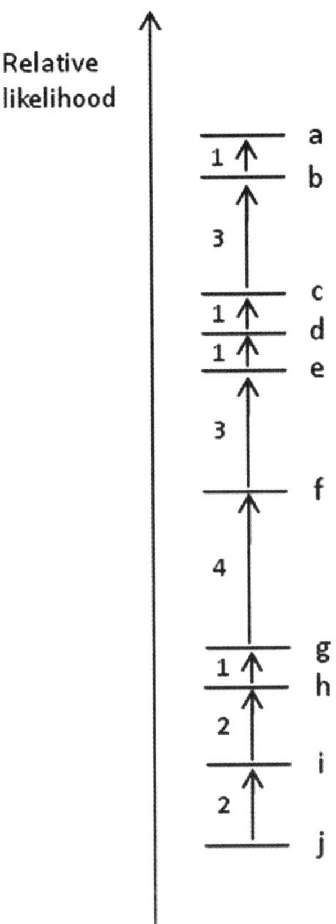

Figure 15.2   A stack of events built from the differences in likelihood between neighbours.

Figure 15.3, supposing that we can only classify things into five broad levels of absolute likelihood. In place of Figure 15.2, we have perhaps the information in Figure 15.4, in which the limited towers we can build have some floors with higher ceilings than others and sit in an unknown landscape, where the ground floors of some may be higher than the roofs of others, but we do not know whether they are or not. The term "tower" is used for these relative reasoning stacks in Meteor, and I will adopt it here also as a convenient term.

The diagrams in Figures 15.3 and 15.4 convey some useful information that is the same in both cases. For example, if the letters in the diagrams represent biotransformations, you will see from Figure 15.3 that if a compound can undergo biotransformations 'a' and 'c', 'a' is more likely to be seen in practice; you can infer the same thing from the information in Figure 15.4, since 'b' is more likely than 'c' and 'a' is more likely than 'b'.

A brief trip to Scotland provides a more concrete illustration of building relative likelihood towers, and highlights a potential pitfall. Consider the

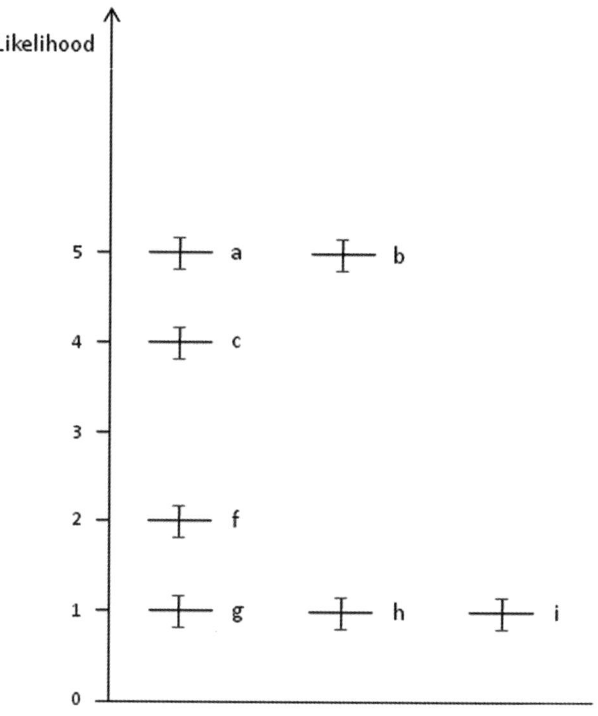

Figure 15.3 Absolute likelihoods of events in a sparsely populated domain.

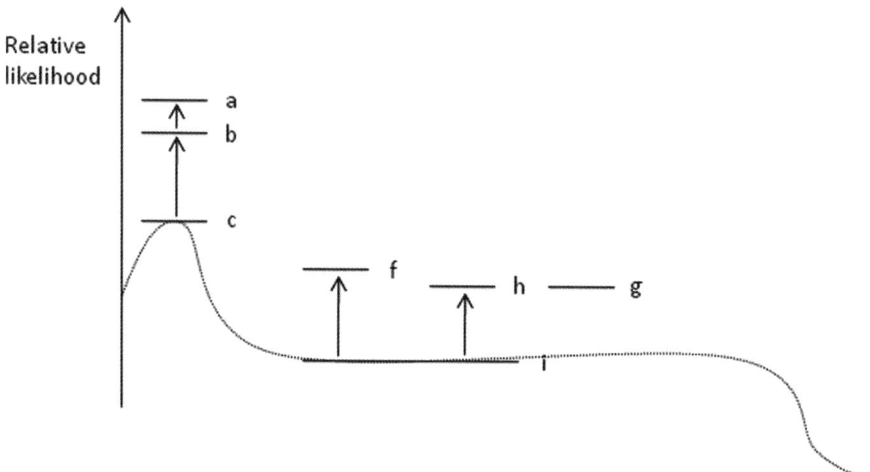

Figure 15.4 A "landscape" containing towers built from what is known about relative likelihood between events.

following statements about conditions on Ben Nevis mountain and in Glen Nevis valley beneath it:

snow in Glen Nevis in December is more likely than in August; (1)

snow in Glen Nevis in February is more likely than in December; (2)

snow on Ben Nevis is more likely than snow in Glen Nevis in the same month. (3)

From statements (1) and (2) we can predict that snow in Glen Nevis in February is more likely than in August. Taking (3) into consideration as well we can predict that snow on Ben Nevis in February is more likely than snow in Glen Nevis in August. But beware! Snow on Ben Nevis in February is not necessarily more likely than snow on Ben Nevis in August on the basis of this information. We do not know how much more likely it is to snow on Ben Nevis than in Glen Nevis and whether the difference is the same in every month.

Returning to Figures 15.3 and 15.4, each provides information that the other does not. Figure 15.3 shows 'c' to be at a higher level of absolute likelihood than 'f'. It follows logically that 'c' is more likely than 'f' – a piece of information that was apparently not available in the data from which Figure 15.4 was constructed. Figure 15.3 shows both 'a' and 'b' to be more likely than 'c', in absolute terms, but cannot distinguish between 'a' and 'b', while Figure 15.4 shows that 'a' is more likely than 'b'. Figure 15.3 shows 'g', 'h', and 'i' all to be at the same broad level of absolute likelihood. Figure 15.4 shows that 'h' is more likely than 'i' and that 'g' is as likely as 'h' (and thus that 'g' is also more likely than 'i').

Building relative likelihood towers relies on the assumption that the domain in which the rules were created and the current one are the same. Biotransformation 'a' obviously would not be more likely than biotransformation 'c' in a cell that contained none of the enzyme that promoted biotransformation 'a'. But it would be impossibly complicated to write conditional rules of the form "If xyz is true then 'a' is more likely than 'c'" covering all the possible combinations of biotransformations, cell types, and species, and it would defeat the idea of letting the program build relative reasoning stacks automatically (*i.e.* letting it find the relationship between 'a' and 'c' when given only rules about 'a' and 'b', and 'b' and 'c'). Applying absolute reasoning *prior* to relative reasoning is a way of dealing with the problem. Absolute rules might say that if the species of interest is a mammal, then biotransformation 'a' is probable, but if the species of interest is a bacterium then biotransformation is doubted, and they might say that 'c' is probable in both species. If the user asks about bacteria, 'a' will be classed as doubted by absolute reasoning and 'c' will be classed as probable. In a hierarchy in which absolute reasoning takes precedence, and relative reasoning is used only to discriminate between predictions having the same approximate absolute likelihood, the relative reasoning rules linking 'a' and 'c' will not be activated.

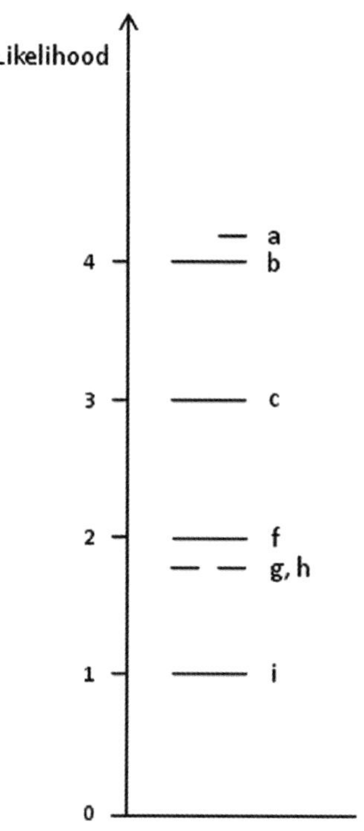

Figure 15.5   A way of combining information about absolute and relative likelihood.

Grouping biotransformations broadly into absolute levels of likelihood brings together the information from Figures 15.3 and 15.4 into a single diagram, reminiscent of a quantum energy diagram, such as the one in Figure 15.5. Perhaps, for example, 'g', 'h', and 'i' are all plausible. The definition of 'plausible' used in Meteor is the same as in Derek for Windows (see Chapter 13), namely that there are arguments for and against but the balance of the evidence is for, and the definition of 'probable' is that there are arguments for and none against. So all things that are plausible need not necessarily be equally likely, and neither need all things that are probable. Figure 15.5 represents this idea, where 'g', 'h', and 'i' are all in band 1 but 'g' and 'h' are both more likely than 'i', and 'g' is as likely as 'h'. There remain many unknowns. The reality might be more like what is represented in Figure 15.6: we only know that 'g' and 'h' are approximately equally likely and we do not know how much more likely they are than 'i'; absolute levels may be clearly defined, but if it were possible to attach numerical probabilities to them (which it is not) it is unlikely that they would turn out to be equally spaced they way they are drawn in Figure 15.5. However, as long as the assumption holds that we can classify biotransformations broadly into bands of absolute likelihood, Figure 15.5 is a sufficient representation of the available knowledge for us to see that, for example, 'a' is the most likely thing of all; 'c' is more likely than any of 'f', 'g',

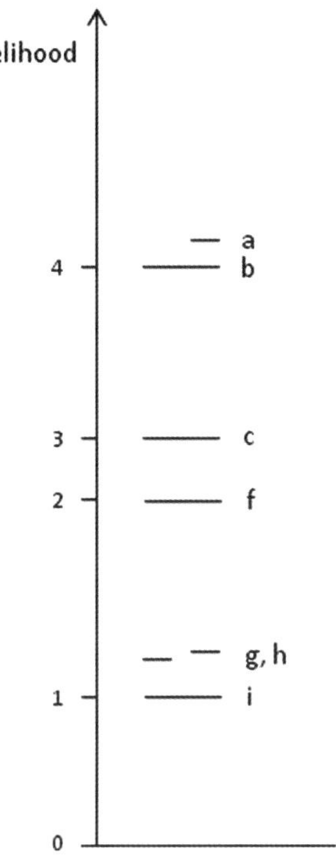

Figure 15.6 How Figure 15.5 might look, depending on the sizes of relative likelihood differences.

'h' and 'i'; 'g' and 'h' are about equally likely within the limits of accuracy of the model, and both are more likely than 'i'.

Meteor uses this two step approach, giving absolute reasoning priority over relative reasoning, although, for reasons of history, neither the way that it was first implemented and reported[1], nor the way it currently works, assures this prioritisation 100%. An assessment of absolute likelihood is associated with each biotransformation in the knowledge base, taking into consideration structural and physico-chemical factors that may influence binding to an enzyme site or the ease with which a compound can reach the site. A separate set of rules describes the relative likelihood that one biotransformation will win out over another if they are in competition. A pair of biotransformations covered by a relative rule may describe the same reaction promoted by the same enzyme, but for reaction centres in different environments, for example, hydroxylation of a benzylic methylene group compared with oxidation of a terminal methyl group. Alternatively, a pair of biotransformations may be for reactions that are different, but which compete for similar reaction sites – for example *N*-methylation and *N*-acetylation of a primary amine. Knowledge base writers enter relative rules of the two kinds mentioned earlier in this chapter: "A" is more likely than "B" and "A" is as likely as "B". Of course, they enter them only for cases where information is available, which by no means covers all the biotransformations in the knowledge base. Knowledge base writers do not need to work out the implied relationships such as that if 'a' is more likely than 'b' and 'b' is more likely than 'c', then 'a' is more likely than 'c'. Meteor constructs for itself all the towers that can be constructed.

A Meteor user can specify an absolute reasoning cut-off value below which biotransformations will not be applied, and can also specify how many levels from the tops of the relative reasoning towers are to be retained. By default, the absolute reasoning cut-off value is "plausible" and the number of relative reasoning levels retained is 1. With these settings, at each level in the metabolic tree Meteor discards biotransformations that are less than plausible. Where more than one biotransformation in the same relative reasoning tower can be applied to the query structure or metabolite currently under consideration, Meteor retains only the one highest in the tower.

These were the constraints set when the output from Meteor shown in Figure 14.6 was generated. The " $\geq$ "symbol at the top right of the box for biotransformation 73 indicates that it has been retained in preference to one or more other biotransformations that are relatively less likely. The relative reasoning report, slightly modified and reproduced in part in Figure 15.7 represents the relative reasoning tower in the form:

$$\mathbf{73} = 65$$
$$\mathbf{78}$$

showing 73 and 78 in bold type because they are the biotransformations relevant to the structure being processed. Biotransformation 73 has been chosen in preference to biotransformation 78, because biotransformations 73 and 65 are equally likely and biotransformation 65 is more likely than biotransformation 78. If the query is reprocessed with the relative reasoning level set to 2 instead of 1, a larger metabolic tree is generated (Figure 15.8), and Figure 15.9 shows the part of it relevant to this discussion. Biotransformations 73 and 78 now both appear.

Using a combination of absolute and relative reasoning, Meteor generates metabolic trees which are close enough to what is found in practice to make the program useful. The main metabolites found experimentally for many compounds are predicted by Meteor, but if the reasoning cut-offs are set lower to find the metabolites that are otherwise missed, Meteor usually appears to overpredict. This is a difficult area in which to conduct evaluations, because finding a structure in a Meteor tree that was not reported experimentally will often not mean that Meteor was wrong. It is not unusual for a metabolite to be formed, but not reported in an experimental study – either because only a small amount of it was formed; because of limitations in procedures for isolating or detecting metabolites; or because the only purpose of the experiment was to look for some other metabolite.

Parent = Submitted 72
Relative Reasoning level = 1

Biotransformation 73 was retained in preference to 78.

<u>Relative Reasoning Rules</u>

73 = 65
78

65 = 73
Relative000312
Hydroxylation of Methyl Carbon Adjacent to an Aliphatic Ring = Hydroxylation of
Terminal Methyl
Comments: Neither of these two aliphatic hydroxylations is more favoured
sterically or electronically and so they are considered equally likely.

65 > 78
Relative0000324
Hydroxylation of Methyl Carbon Adjacent to an Aliphatic Ring > Para Hydroxylation
of Monosubstituted Benzene Compounds
Comments: Taking electronic and steric requirements into consideration,
biotransformation 65 is considered more likely than biotransformation 78.

Figure 15.7   A relative reasoning report from Meteor.

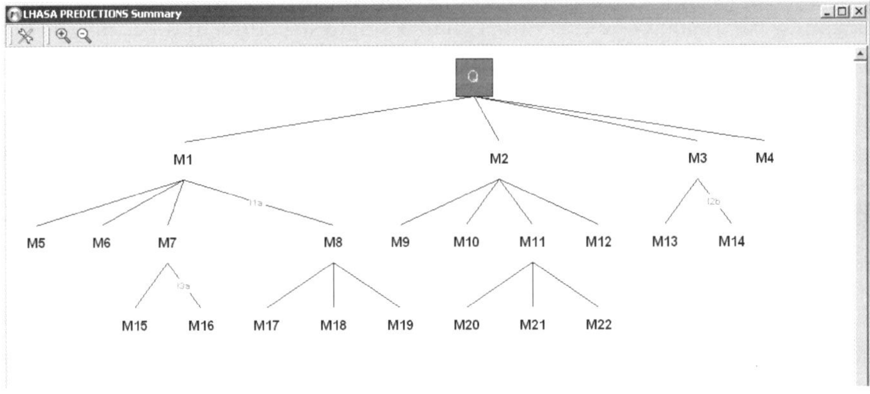

Figure 15.8   The meteor metabolic tree for ethyl benzoate when relative reasoning is
set at level 2.

It may have crossed your mind during the course of reading the preceding
chapter and this one, that using reasoning is not as simple as you might have
hoped. One of the reasons for the popularity of Derek for Windows is that it is
easy to use. It thinks like a toxicologist, so to speak, and you can easily explore
its reasoning and understand how it has reached its conclusions. Meteor may be
less easy to understand, both because of the inherent, greater complexities of

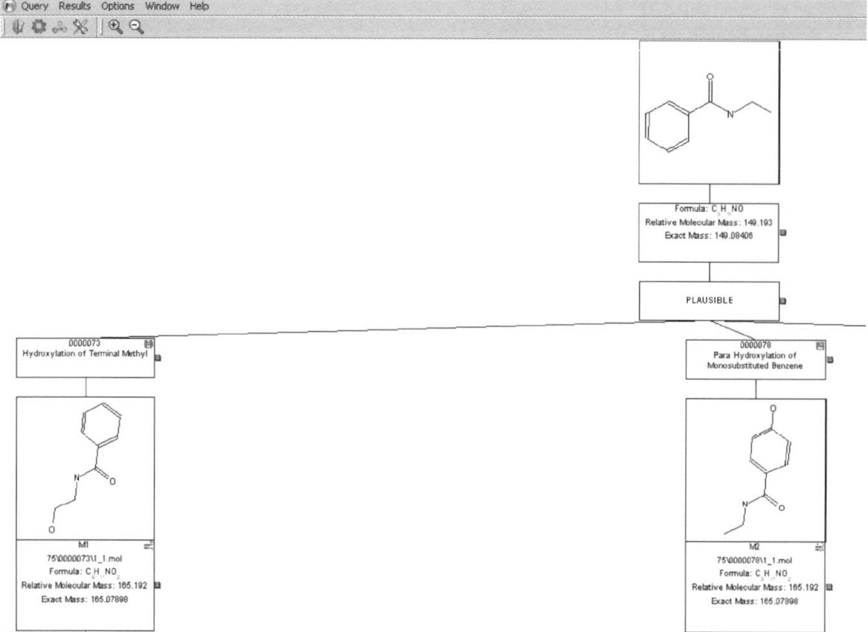

Figure 15.9   Biotransformation 78 now appears on the tree.

dealing with a metabolic tree rather than a single structure, and because of the combined use of absolute and relative reasoning. It is not obvious to a novice user from the way the output is displayed, that less likely metabolites may have been generated but discarded, and reasoning reports are not easy to follow unless you understand how Meteor works. The underlying ideas of absolute and relative reasoning are straightforward: using both, each with its own ramifications when the interactions between those underlying ideas are taken into account, is complicated. For the reasoning methods described here to be widely adopted, the challenge will be to design applications which explain themselves effectively and do not require users to be trained in reasoning theory and jargon (or, indeed, even to be interested in it at a theoretical level).

Reference

1. W. G. Button, P. N. Judson, A. Long and J. D. Vessey, Using Absolute and Relative Reasoning in the Prediction of the Potential Metabolism of Xenobiotics, *J. Chem. Inf. Comput. Sci.*, 2003, **43**, 1371–1377.

CHAPTER 16
Predicting Biodegradation

Mammalian metabolism and microbial catabolism – the biological process of greatest interest in connection with environmental degradation of chemicals – are different both in their biological purpose and their consequences. Xenobiotic chemicals may be metabolised to make them easier to excrete or, if they are mistakenly taken for something else by an enzyme, in an attempt to utilise them for a specific purpose. The purpose of microbial catabolism is to generate energy – the chemical is being used as food. (The degradation of chemicals in mammalian systems can actually include an element of bacterial catabolism as well as metabolism, because of the actions of bacteria in the gut). In the presence of oxygen, the ultimate catabolic end-product from the carbon content of organic compounds is carbon dioxide. So the rate of degradation of a chemical in soil or water is typically expressed in terms of the rate at which it is converted to carbon dioxide or, for practical reasons, more usually the rate of consumption of oxygen is measured (the biological oxygen demand, or BOD).

Ever since the environmental effects of pesticides and early detergents began to raise concerns[1,2] there has been interest in the ease of degradation of chemicals in the environment. To be more strict, the concern has been about persistence, but that comes down to the same thing – it is just a matter of whether you ask how easily something degrades or how reluctantly. Until recently, the question has usually been only whether a chemical is broken down quickly or is likely to persist. Provided that the half-life of the chemical is short, there may be no reason for particular concern, even if the chemical is toxic to some kinds of wild life. In practice, it will normally be highly diluted in the environment because release will be balanced by degradation – except temporarily and locally where, for example, there is an accidental spillage. Increasing concerns about the very strong effects of some chemicals in the environment – for example, the suspected effects of endocrine disruptors[3] – have led to greater interest in what the degradants of chemicals are, and not just

RSC Theoretical and Computational Chemistry Series No. 1
Knowledge-based Expert Systems in Chemistry: Not Counting on Computers
By Philip Judson

Published by the Royal Society of Chemistry, www.rsc.org

in the rate of degradation. The European REACH legislation will require companies to provide more information than in the past, about the actual or predicted degradants of chemicals in the environment.

Joanna Javorska, Robert Boethling and Philip Howard wrote a very informative paper about methods for predicting biodegradation in 2003.[4] I limit this chapter to the well-known examples of knowledge-based systems.

16.1   BESS

BESS[5] is a knowledge-based system for predicting biodegradation. Its primary goal is to answer the question: "Is my compound [readily] biodegradable?", but it generates trees containing the structures of the biodegradants it predicts. Rules in the knowledge base describing biodegradation reactions are written in Smalltalk™, an object-oriented programming language. Each rule contains a description, in the form of a linear textual code, of the substructure that keys the reaction and the corresponding substructure that is generated in the product. For example, $Rn\text{-}CH\text{=}CH\text{-}Rm!Rn\text{-}CH\text{-}CHOH\text{-}Rm$ n; $m > 0$ represents the addition of water to an olefin in which the atoms adjoining each end of the double bond are carbon atoms. The reaction descriptions are classed according to the conditions under which they occur. They are organised into groups within which reactions normally inter-operate to produce some overall result. For example, a group of rules operates sequentially and repeatedly to remove terminal two-carbon fragments from chains by the β-oxidation process. An editor allows knowledge base developers to add reactions by completing fields for keying substructures, product substructures, conditions, references, *etc.* without needing to write code in Smalltalk™.

Presented with a query, BESS first checks in a database of published biodegradation pathways to find out if the fate of the query is already known. If so, it is reported to the user. If not, BESS applies rules from the knowledge base that are appropriate to the conditions under consideration to generate first level degradation products. It checks to see if any of these are included in the database of published pathways and applies rules from the knowledge base to them if not, and so on, until the possibilities are exhausted. The user can opt simply to be informed whether, at the end of the process, the chemical has been fully mineralised (*i.e.* converted to carbon dioxide and simple heteroatom-containing compounds such as ammonia and water), but he/she can also explore the degradation tree, select structures from it and process them under different constraints.

The team who developed BESS have done some research into making the program able to learn from data, for example by using genetic algorithms.

The rules about reaction conditions impose some control over the size of the degradation tree, but BESS does not otherwise attempt to discriminate between more and less likely reactions, or major and minor products. One of the difficulties with predicting biodegradation is that it is much more indiscriminate than metabolism anyway. To some extent, if a reaction can happen, a bacterium will turn up to do the job and it will happen. Environmental conditions, in soil for example, are so variable that it is currently impossible even to extrapolate reliably from laboratory experiments to field situations, let alone to make firm predictions from computer models. There is an argument for saying that predicting everything that might happen is the right thing to do, complicated though it makes things.

16.2 CATABOL

Another knowledge-based system is CATABOL,[6,7] which continues to be developed and supported and is widely known. CATABOL uses a knowledge base of biodegradation and abiotic degradation reactions with which probabilities of occurrence are associated, based on assessment by experts of published data. It is assumed that there is a relationship between the observed probabilities of biodegradative reactions and their rates and so, having assessed what degradation reactions are likely, CATABOL calculates from their probabilities the expected theoretical BOD in one of the standard tests described by the OECD[8], and it is reported to the user (the maximum theoretical BOD is the amount of oxygen that would be consumed were all of the carbon in the chemical to be converted into carbon dioxide but this may not be achieved in a standard test, which is run for a prescribed length of time).

CATABOL takes a depth-first approach. That is, it starts by looking for, and reporting, the single degradation pathway most likely to predominate, as distinct from generating first the products of all likely single step reactions – the breadth-first approach.

The assumption that observed probabilities, reaction rates, and hence product quantities, are related is not necessarily true, but it is pragmatic. What catabolic products are seen in practice in an experiment depends very much, though not exclusively, on the competitive success of different catabolic reactions which in turn can be expected to depend on their relative rates. So one can hope that, when averaged out over many studies, the probabilities of observing reactions (*i.e.* for each reaction, the ratio of the number of studies in which a reaction is observed to the number of studies in which the keying substructure is present in a test structure or a degradant) will reflect their relative rates. Given the relative reaction rates, you can estimate the relative quantities of degradants which will be formed. There is contention among scientists working on the prediction of biodegradation, about whether these are acceptable assumptions and you may want to form your own view.

A metabolism knowledge base has been added to CATABOL and so it can also be used to predict mammalian metabolism.

16.3 The UMBBD, PPS and Mepps

Teams led by Lynda Ellis and Larry Wackett at the University of Minnesota in Minneapolis and St. Paul have developed a database of known biodegradation pathways, the University of Minnesota Biocatalysis/Biodegradation Database (UMBBD),[9] which is accessible free of charge *via* the worldwide web.[10] Drawing on the information contained in the UMBBD, they developed a web service for predicting biodegradation reactions, the Pathway Prediction system (PPS).[11,12] Like BESS, the system generated all possible first level products by applying rules describing biodegradation reactions to the query compound, but to avoid exponential growth of the biodegradation tree, it did not go on automatically to process those products further. The products were displayed to the user who could then make selections for further processing to create the next level of biodegradants. Thus routes to complete mineralisation could be constructed under the control of the user.

Letting the user decide which branches of the growing biodegration tree to extend is an effective way of controlling the ultimate size of the tree but it is, of course, subjective. So the team at the University of Minnesota set up a workshop to discuss the feasibility of predicting how likely biodegradation reactions would be. I was one of the attendees. It was the second of what became a series of four workshops to date,[13] the first having been on the creation of PPS and the later ones continuing discussions both on the development of PPS, and ways to improve understanding of the factors determining which biodegradation reactions are seen in practice. Experts on microbial environmental degradation had been asked in advance to assess how likely the biodegradation reactions covered by PPS were, in broad terms – very likely, likely, neutral, unlikely, or very unlikely. The reactions had been divided into blocks so that each reaction was considered by external experts and, in addition, Larry Wackett looked at all of them. Each set was reviewed by at least two experts.

There was agreement about the likelihood of many reactions, which was encouraging. Human experts felt able to make the predictions and it might be hoped that their being in agreement was a sign that their predictions had some basis in fact. Most of the two days was spent in discussion about the areas of disagreement. Minor disagreements, such as that one person assessed a reaction to be "very likely" while two others assessed it only to be "likely", were not seen as problems given the uncertainties of the task. Major disagreements, such as cases where two people thought a reaction was "very likely" and one thought it "very unlikely" could mostly be resolved through discussion: inevitably in a project on this scale, one or two were simply the result of mistakes and quickly corrected; in some cases, one individual had specialised knowledge which swayed the views of the others leading to agreement; in others, one individual's views had been biased by unusual biodegradation in a specialised area and again agreement was reached. There remained cases where it was not possible to reach agreement because, as the discussions revealed, either the available information really was contradictory or there was not enough of it to make an informed judgement.

The overall conclusion was that, while there were many gaps, enough was known for it to be worth trying to incorporate information about the likelihood of reactions into a biodegradation prediction system. The team at the University of Minnesota collaborated with, and were sponsored by, Lhasa Limited to work on the idea. A sister system to PPS, to run under Microsoft Windows on PCs, was developed using the Meteor technology and given the name "Mepps", for "Meteor environmental pathway prediction system". As rules about the absolute likelihood of reactions were added to Mepps they were also added in parallel to PPS. In place of the likelihood terms used in Meteor, Mepps and PPS use the terms "very likely", "likely", "neutral", "unlikely" and "very unlikely", and colour coding is used to show the ranking of each product. The evolution of thinking in the project paralleled what happened in the Meteor project: the need to consider relative likelihood became apparent and so

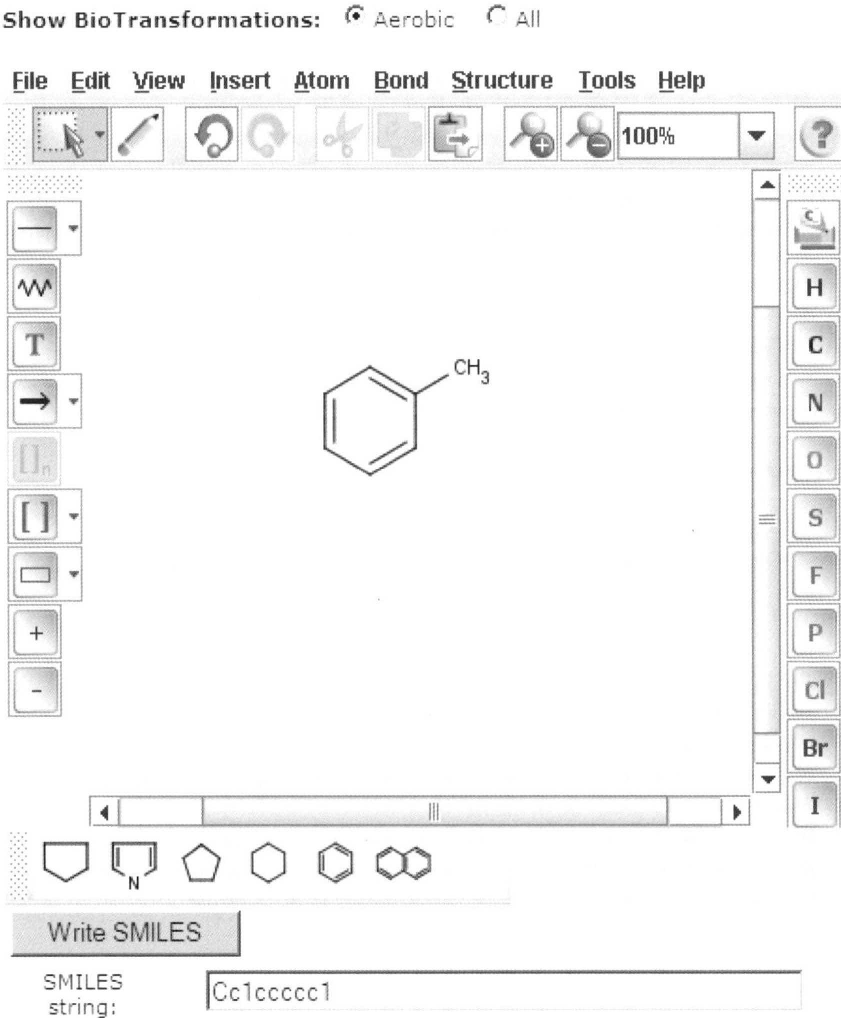

Figure 16.1   Using MarvinSketch to draw a structure into PPS.

relative reasoning rules have also been developed[14,15] At the time of writing of this book, only some of the relative reasoning rules have been added to Mepps.

When you use PPS online you can enter a SMILES string for the structure you want to process or you can draw it using MarvinSketch,[16] (which downloads to your computer from the PPS site) and have the SMILES string generated automatically. Figure 16.1 shows toluene entered for processing as a structure in MarvinSketch. You can specify whether you want to see only aerobic biodegradation reactions or both aerobic and anaerobic; whether you want processing to stop at a single step or want it to continue to a second step; and whether you want to see all predictions or only those that are at least of neutral likelihood.

With the option set to see only a single step, the output for toluene is as shown in Figure 16.2. A colour key at the top of the page shows the interpretation of the coloured bars beside the reaction products. In this example, there are no unlikely or very unlikely reactions/products, but if there were they would be identified by differently shaded bars. The use of relative reasoning is not apparent, but may have influenced what has been retained for display. You can select a product and request further processing. The ranking of the reactions may help you to decide which appears to be the most important (or, in this case, the more important, there being only two), but you are not restricted

Aerobic Likelihood:
Very likely
Likely
Neutral
Show BioTransformations:
Aerobic
All
Choose the next reaction step:
bt0353
Next
Rule
Cpd
CH3
HO
HO
bt0036
Next
Rule
Cpd
HO

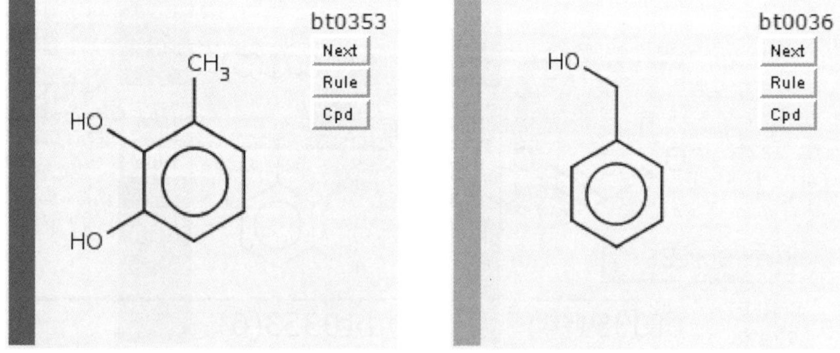

Figure 16.2 Output from PPS for a single biodegradation step.

to doing so. With the option set to see two steps, the output is as shown in Figure 16.3.

While the ranking is actually of the reactions, it is the products that are labelled. In this context it makes no difference, but note that the distinction is important. The likelihood of a reaction determines the likelihood of formation of its products by that route. If, however, you want to know how likely it is that you will find the product if you analyse a sample from a live experiment, this may not be the right answer. The product labelled bt0353(1) in Figure 16.3 is likely to be formed, but it is also likely to be converted into the products labelled bt0351(3), bt0351(4) , bt0351(5) and bt0351(6), and so you may find

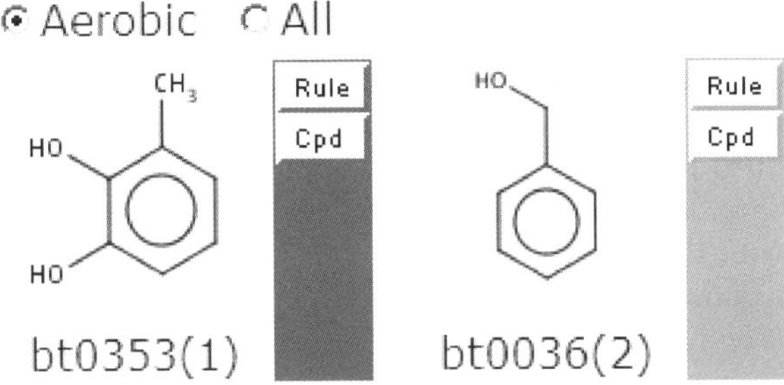

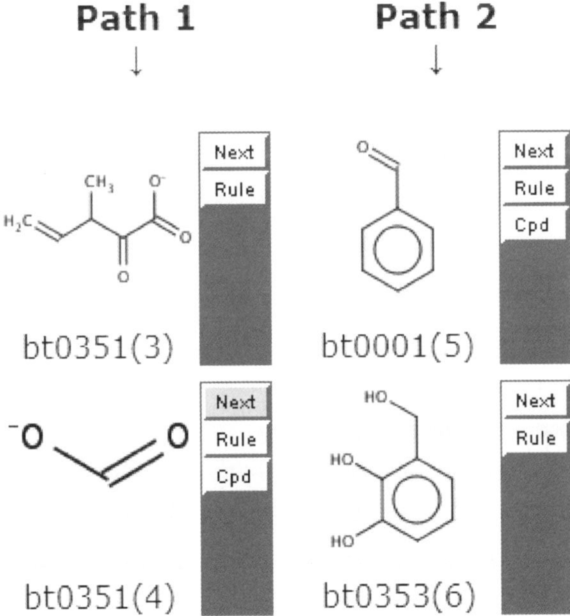

Figure 16.3 Output from PPS for two biodegradation steps.

very little of it in practice. The products labelled bt0001(7) is flagged as likely because the reaction leading to it is likely to take place, provided that the reaction precursor is available, but the likelihood of formation of the precursor in this case, bt0036(2), is neutral. So the likelihood of actually finding bt0353(6) cannot be greater than that. Or, at least, not in this example – in a more complicated case there might be more than one route leading to the same product, each displayed in a different part of the tree. To determine the likelihood of finding the product, you would need to consider the likelihood of every reaction leading to it.

Mepps shows much the same information, in a different style, but user control of the degradation tree is the same as in Meteor. You can request that processing be stopped at any number of steps, rather than one or two, although this is not a very exciting difference. For the PPS online service, stopping at Step 2 makes sense because of the delays that there would be in communication if the tree were allowed to grow too much. For PC-based Mepps, it would be nice to think that the program could be allowed to keep running until all possibilities at, or above, the chosen cut-off level for likelihood had been explored. But with the current amount of knowledge in the program about the likelihood of reactions, an open-ended search can lead to trees that are too large to be built within a reasonable time, or take up too much memory. So it is wise to put a limit on the number of steps. The default is six.

To illustrate Mepps, processing atrazine with absolute reasoning set to neutral or greater and relative reasoning limited to the most likely reactions, leads to the biodegradation tree shown in Figure 16.4. In both Meteor and Mepps, processing stops when a structure is generated that has already been generated and processed earlier in the tree, which is why, for example, the second instance of amino-3-ethylamino-5-hydroxy-s-triazine is not processed further. The knowledge base of Mepps version 10.0.2, the version used to generate the tree for this illustration, does not contain reaction descriptions for further degradation of the amino- and hydroxy-substituted triazines which therefore terminate the sequences in the tree.

16.4 META

The META program, for predicting metabolism, was mentioned in Chapter 14.2. It also contains knowledge modules for predicting biodegradation discovered by using the Multicase system for analysing data (see Chapter 11.3).[17] The researchers were able to find some rules for anaerobic biodegradation, which is more difficult to work on than aerobic biodegradation because of a lack of experimental data.[18] An additional feature of META is a photodegradation module, to predict abiotic reactions promoted by sunlight in the environment.[19] The title of the paper about it refers to "natural-like" reactions, because the information that the researchers used for evaluation of their model came mainly from laboratory studies in dilute aqueous solution under UV light, rather than from field studies. Something over a dozen types of photocatalysed reactions and

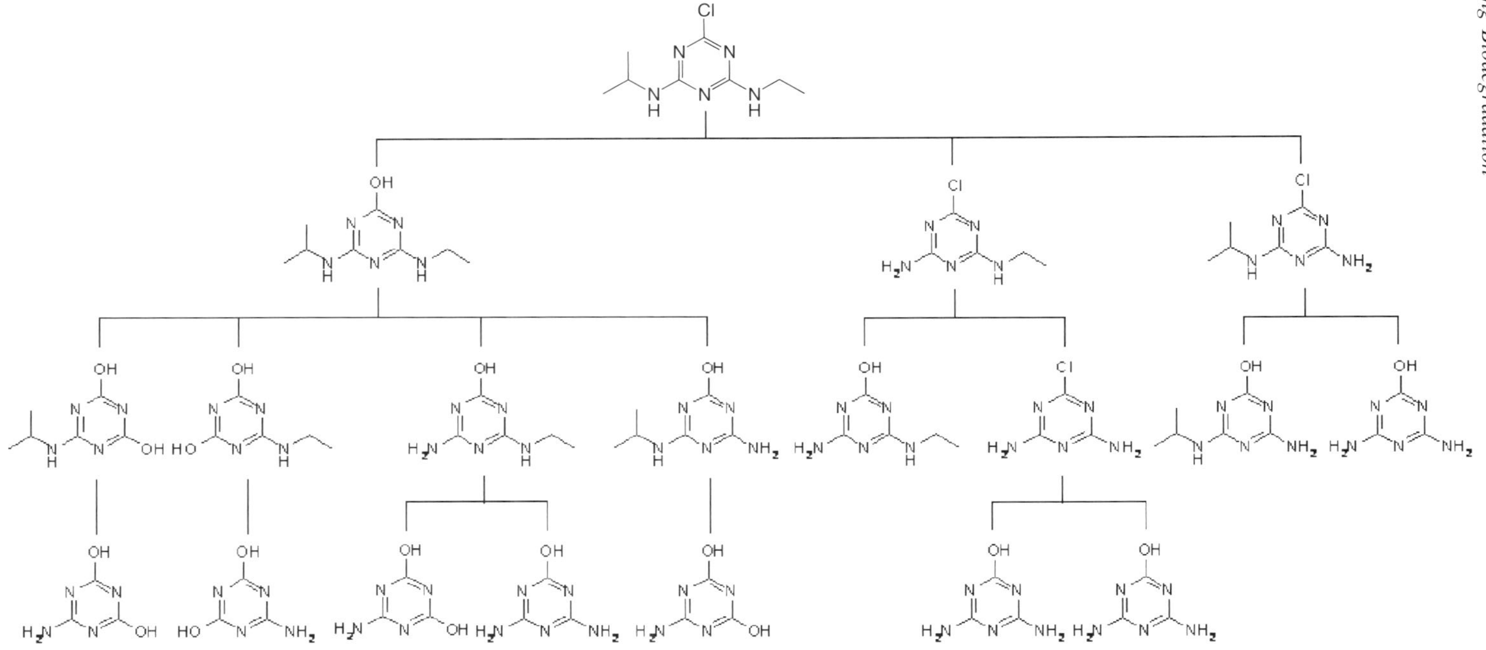

Figure 16.4   A Mepps biodegradation tree for atrazine.

chains of further spontaneous reactions that commonly ensue, are covered by several hundred reaction descriptions.

16.5   The Future for Prediction of Environmental Degradation

CATABOL has some success in predicting biodegradation rates and pathways at its current stage of development and work continues on its improvement. Mepps is at about the same stage of development as Meteor was when it was first made available as an experimental system. Big improvements have been made to the prediction of metabolism by Meteor since then, and it can be hoped that the performance of Mepps will improve similarly. However, the variability of environmental conditions and the versatility of bacteria make it very difficult to predict environmental degradation pathways. It is unlikely that any system will be able to do better than to give broad indications. Whether that will be enough to make them widely useful is not clear.

Funding for the development of prediction systems for toxicity and mammalian metabolism has come largely, though not exclusively, from the pharmaceutical and agrochemical research communities. With one or two important exceptions, the chemical manufacturing industries do not traditionally sponsor this kind of research. While environmental issues are important to the agrochemical and chemical manufacturing industries, they have so far concerned the pharmaceutical industry much less. Agrochemical companies cannot be expected to fund research into the prediction of biodegradation pathways on their own. Where will the impetus and the funding come from? Will the introduction of REACH, alongside generally increasing concern about environmental issues, be enough to carry these projects forward?

References

1. R. Carsons, *Silent Spring*, Houghton Mifflin, Boston, Massachusetts, 1962.
2. R. D. Swisher, *Surfactant Biodegradation*, Marcel Dekker, New York, 1987.
3. *Global Assessment of the state-of-the-science of endocrine disruptors*, ed. T. Damstra, S. Barlow, A. Bergman, R. Kavlock and G. Van Der Kraak, World Health Organisation, 2002.
4. J. S. Jaworska, R. S. Boethling and P. H. Howard, Recent Developments in Broadly Applicable Structure-Biodegradability Relationships, *Environ. Toxicol. Chem.*, 2003, **22**, 1710–1723.
5. W. F. Punch, A. Patton, K. Wight, R. J. Larson, P. H. Masschelen and L. Forney, A Biodegradability Evaluation and Simulation System (BESS) Based on Knowledge of Biodegradation Pathways, in *Biodegradability Prediction*, ed. W. J. G. M. Peijnenburg and J. Damborsky, Kluwer, Dordrecht, 1996, pp. 65–73.

6. J. Jaworska, S. Dimitrov, N. Nikolova and O. Mekenyan, Probabilistic Assessment of Biodegradability Based on Metabolic Pathways: CATABOL System, *SAR QSAR Environ. Res.*, 2002, **13**, 307–323.

7. CATABOL is supplied by the Laboratory of Mathematical Chemistry, University "Prof. Assen Zlatarov", 1 Yakimov Street, Bourgas, 8010 Bulgaria.

8. *OECD Test Guidelines, Section 3: Degradation and Accumulation*, Organisation for Economic Co-operation and Development, Paris, 2005.

9. L. B. M. Ellis, D. Roe and L. P. Wackett, The University of Minnesota Biocatalysis/Biodegradation Database: the First Decade, *Nucleic Acids Res.*, 2006, **34**, D517–D521.

10. http://umbbd.msi.umn.edu/

11. B. Kyeng Hou, L. P. Wackett and L. B. M. Ellis, Predicting Microbial Catabolism: a Functional Group Approach, *J. Chem. Inf. Comput. Sci.*, 2003, **43**, 1051–1057.

12. http://umbbd.msi.umn.edu/predict/

13. Information about the workshops can be found on the University of Minnesota web site at http://umbbd.msi.umn.edu/predictbt/index.html.

14. K. Fenner, J. Gao, S. Kramer, L. B. M. Ellis and L. P. Wackett, Data-Driven Extraction of Relative Reasoning Rules to Limit Combinatorial Explosion in Biodegradation Pathway Prediction, *Bioinf.*, 2008, **24**, 2079–2085.

15. L. B. M. Ellis, J. Gao, K. Fenner and L. P. Wackett, The University of Minnesota Pathway Prediction System: Predicting Metabolic Logic, *Nucleic Acids Res.*, 2008, **36**, W427–W432.

16. MarvinSketch comes from ChemAxon Kft., Máramaros köz 3/a, Budapest, 1037 Hungary.

17. G. Klopman and M. Tu, Structure-Biodegradability Study and Computer-Automated Prediction of Aerobic Biodegradation of Chemicals, *Environ. Toxicol. Chem.*, 1997, **16**, 1829–1835.

18. E. Rorije, W. J. G. M. Peijnenburg and G. Klopman, Structural Requirements for Anaerobic Biodegradation of Organic Chemicals: a Fragment Model Analysis, *Environ. Toxicol. Chem.*, 1998, **17**, 1943–1950.

19. A. Sedykh, R. Saiakhov and G. Klopman, META V. A Model of Photodegradation for the Prediction of Photoproducts of Chemicals under Natural-Like Conditions, *Chemosphere*, 2001, **45**, 971–981.

CHAPTER 17

Other Applications and Potential Applications of Knowledge-Based Prediction in Chemistry

17.1 The Maillard Reaction

I spent a couple of years at Radboud University in Nijmegen looking at the use of a modified version of LHASA to predict the so-called Maillard reaction. The Maillard reaction is the process that accompanies browning of food during baking, among other things. It is not one reaction, but an ensemble of reactions between sugars and aminoacids for which the "Maillard Process" might be a better description.[1] It generates so many products that food chemists are unable to predict the formation of undesirable ones with unpleasant flavours. Researchers at the company sponsoring our research were conducting experiments in which they heated an aqueous solution of a single sugar with a single aminoacid and buffering agents to control pH for varying periods, and then identified the main components of the resultant mixtures using GC-MS (gas chromatography linked to mass spectrometry).

We wrote a special knowledge base for LHASA of about twenty reaction descriptions to cover the main reactions of sugars with themselves and with aminoacids. The potential complexity of the Maillard process is astonishing. With an open-ended search, the program generated reaction trees for a single sugar and a single aminoacid containing different products which could be counted in millions. Among them were all sorts of structures, including heterocycles. If the amino acid was cysteine, having a thiol group in its structure, the total number of products increased dramatically and, not surprisingly, included a variety of sulfur heterocycles. Applying constraints to take account of likely changes in concentration, brought the numbers down to hundreds: as

RSC Theoretical and Computational Chemistry Series No. 1
Knowledge-based Expert Systems in Chemistry: Not Counting on Computers
By Philip Judson

Published by the Royal Society of Chemistry, www.rsc.org

time progresses the starting materials and early products will gradually disappear and so reactions between them and much later products might be expected to be less significant; due to the sheer number of products created, those appearing late in the synthesis tree are likely to be created only in small quantities (provided that conditions do not strongly favour only one or two narrow reaction routes).

The practical experiments using GC-MS by researchers at the company sponsoring our work, showed that we were having some success in predicting the major components in reaction mixtures, but those are not the ones food scientists need to have predicted. An expert would expect them anyway. It was more important to predict relatively minor components that might be surprising and that had undesirable odours or flavours. Given the nature of the reaction, there will be a very large number of minor products, and Maillard LHASA predicted many more than a human expert would be likely to think of, or willing to take the time to write down. It was not possible to do the analyses that would be necessary to find out whether the hundreds of predictions made by Maillard LHASA were correct, and there were so many that it was not practical for a human expert to look at them all. Turning the question round and asking Maillard LHASA, "how did we get this compound from this sugar and aminoacid?" was easier to test. For example, Maillard LHASA was able to explain the appearance of large amounts of acetone from one sugar and aminoacid *via* a reaction sequence that was far from obvious, and had not previously been worked out.

The research project came to the end of its funding period and for a variety of reasons there was no follow-up project, but the possibility has been demonstrated. As I mentioned in Chapter 3.1.7, the GRAAL program might also find use in the future for predicting products of the Maillard process.

17.2 Recording Information about Useful Biological Activity

The technology used in knowledge-based systems to predict toxicity on the basis of the presence of toxicological alerts, or toxicophores, is equally suited to storing and retrieving information about pharmacophores and structural features having useful pharmacological properties, and, as mentioned in Chapter 10.1, this was done in **PHARM-MATCH**. Some companies may store information about pharmacophores, or about structural features conferring useful pesticidal activity on chemicals, in systems of their own or in programs supplied for use in toxicology such as Derek for Windows, but little has been published on the subject. Given that all companies interested in developing biologically active compounds need to ensure that their staff have easy access to existing company knowledge, and given how easy it is to edit knowledge bases in programs like Derek for Windows, it is surprising that no-one seems to supply an application adapted specifically for the purpose. There appears to be an opportunity waiting for the right entrepreneur.

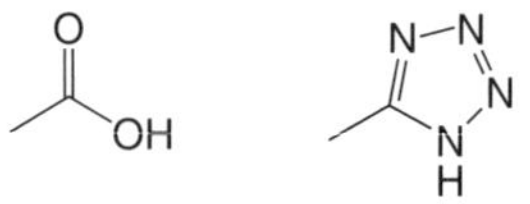

Figure 17.1   A carboxylic acid group and a tetrazole.

17.3   Proposing Structural Analogues for Drug Design

Kent Stewart and colleagues have described a program to help pharmaceutical research chemists to design analogues of pharmacologically active compounds.[2] Some substructural features are known to behave similarly to others in a biological context, for example, tetrazoles can mimic carboxylic acid groups (see Figure 17.1), and ethers and thioethers often have similar biological properties. Their web-based application, "Drug Guru" (derived from "drug generation using rules") contains one hundred and eighty six rules for conversions between substructures having similar biological properties, expressed as SMIRKS strings (see Chapter 5.2). The user enters a structure by drawing it or importing it from a file and the program applies the rules and displays a list of potential analogues.

17.4   Predicting Product Degradation during Storage

During storage, the active ingredients of pharmaceutical products can degrade for a variety of reasons, such as oxidation by atmospheric oxygen and hydrolysis by water in the product formulation or in moist atmospheres. These reactions may be accelerated or promoted by exposure to sunlight, or by high or low pH of the product formulation (which may itself change with time). Zeneth[3,4] predicts potential degradation of active ingredients using the same underlying technology as Meteor and Mepps, the systems for predicting mammalian metabolism and biodegradation described in Chapters 14, 15 and 16, but with a knowledge base of chemical degradation reactions. Extensions to the knowledge base language that had been developed for Meteor, allow statements to be made about conditions which favour or inhibit reactions so that they can be taken into account in reasoning about the likelihood of the reactions.

At the time of writing of this book, Zeneth predictions are restricted to degradation of a single query compound (normally a pharmaceutically active ingredient) by the more important reactions that are driven by factors, such as, the presence of oxygen and/or water, elevated temperature and exposure to ultra-violet light. The user draws the query compound and specifies conditions and the program generates a reaction tree in which the reactions are marked as "very likely", "likely", "neutral", "unlikely" or "very unlikely" on the basis of rules in the knowledge base. If the user specifies different conditions (*e.g.* changing pH from 7.0 to 1.0), the output can change dramatically. At this stage, Zeneth is a prototype too new for thorough evaluation to have been completed, and it is yet to be confirmed that what it predicts corresponds to what is seen in practice. Provided that does prove to be the case, chemists should find it useful to be shown the implications of making small changes to formulations or storage conditions, some of which may come as surprises to all but the most assiduous.

In the longer term, it would be interesting to extend the program to making predictions about interactions between the components of mixtures, *i.e.* to allow structures for a complete formulation to be entered as the query rather than the structure of a single compound. There are no technical reasons why this could not be done, but whether it happens will depend on user needs and cost-benefit considerations. The current sponsors of the project want to know primarily about degradation of active ingredients in isolation, since this better corresponds to the standard stability tests.

Zeneth is currently concerned only with the degradation of active pharmaceutical ingredients during storage. One could imagine using the system in a huge range of applications. Just about every product on the market is subject to some degree of chemical degradation, or improvement, during storage – cosmetics, paints, cheese, beer, wines ... the list is endless. There is an overlap between predicting degradation products during storage and predicting the products of thermal decomposition, and so perhaps the GRAAL program (see Chapter 3.1.7) might move into some of these areas as well.

17.5 Designing Production Synthesis Routes

There is increasing concern in the pharmaceutical industry, in particular, about the potential contamination of products with toxic impurities coming from the manufacturing process. In addition to the natural wish of manufacturers to deliver clean, safe products, there are guidelines in Europe[5] about permissible levels of mutagenic impurities in pharmaceutical products and these are likely to be echoed in USA guidelines.

Bringing together the technologies for predicting toxicity and for predicting the products of reactions makes it possible to build a computer application to help development chemists with the design of production synthesis routes. Unlike the early synthesis planning systems, the purpose of this one would not be the automated design of complete, efficient routes to products. It would be a tool box to allow the chemist to explore his/her ideas, and to seek out routes or methods that minimise the risk of contamination of the final product with harmful substances. For example, the system could help with the ordering of reactions in the synthesis sequence, so that potential or known mutagens needed as reagents or ingredients were used as early as possible in the process, reducing the risk that significant amounts would be present in the final product. The system could help with finding sequences of reactions, such that conditions in later stages destroyed hazardous impurities carried forward from earlier stages. In addition to considering reagents and reactants added deliberately during the manufacturing process, the system could generate the structures of likely by-products at each stage, assess their potential toxicity, and take them into account in the overall plan.

I have put these thoughts to some software developers and people in the pharmaceutical industry, but, as far as I know, there are no projects running yet. Over to you...

17.6   New Approaches to Chemical Synthesis Planning

Having just touched on the subject of production route planning, this is an appropriate place to comment on an approach to helping chemists with synthesis planning at a research level that may turn out to be more appealing than the synthesis design systems described in Chapters 2 and 3. When the early synthesis planning systems were developed, they were the only computer tools available to synthesis chemists. One of the spin-off benefits of the research was the development of chemical reaction database systems, as mentioned in Chapter 3, which were of more everyday use to chemists and quickly displaced the synthesis planning systems. Reaction databases are now very large. This means that users are often overwhelmed with information, and intelligent database searching is needed. Several groups have looked at the idea of creating fingerprints to describe the reactions in a database and using them to find the most suitable methods for making a query structure. Circular fingerprints are grown from the sites of reactions and stored in the database. When a query is entered, the database is searched for reactions containing the same core reaction site, and the ones that are retrieved are ranked according to the size of the shell surrounding the reaction site that is identical in the query and the retrieved structure. The results are presented to the user in order of the ranking. This improves the chances of finding, say, the most appropriate publication about an ester hydrolysis quickly, instead of obliging chemists to scan through dozens of them. I am aware of several projects on systems that work in this way that have run, or are running, but they have not been publicised widely and have not yet delivered fully-marketed products. Developers may be reluctant to make the software public, until its benefits are evident to potential users. That depends, among other things, on providing access to large databases which, in turn, will depend on contractual arrangements with the copyright owners of large databases.

17.7   Predicting Ecotoxicity

QSAR methods have been used with variable success for the prediction of environmental toxicity and fate.[6] There are programs for estimating the narcotic toxicity of a chemical to fish and/or daphnia, of which the most well-known is ECOSAR from the US Environmental Protection Agency.[7] A system for predicting aquatic toxicity which applied rules about the relationship between linear solvation energy and acute toxicity, and estimated on the basis of substructural features in query compounds, was described in a book which also covers the use of expert systems in other environmental areas,[8] but until recently there does not seem to have been much support for work on the use of knowledge-based systems to predict ecotoxicity. Narcosis is caused by disruption of membranes by chemicals and it can be modelled rather well from physicochemical properties – quantitatively for groups of related compounds. However, narcosis is, in a sense, a kind of background toxicity. Just as in the case of mammalian toxicity, compounds containing toxicophores that interact specifically with proteins are dramatically more toxic. Gerrit Schüürmann's group in Leipzig have looked at methods for associating atom centred fragments with this "excess toxicity" in fish and daphnia, and using them to predict it for novel structures (see Chapter 11.4.2). Using information about the ACFs in a query structure, and with estimates of its physical properties, such as water solubility, octanol–water partition coefficient and pK_a, calculated by their ChemProp software, an ecotoxicity program can automatically advise on potential excess toxicity, if it could arise, as well as on the likely narcotic strength of the query structure.[9–11]

17.8 Using Knowledge-Based Systems for Teaching

One or two universities use Derek for Windows to support teaching about toxicology, and some companies encourage staff to explore Derek for Windows in order to learn about the alert-based approach to understanding toxicity, but they simply use the program as it stands. As long ago as the late 1970s and early 1980s, Robert Stolow and Leo Joncas experimented with a version of the LHASA program modified for teaching purposes, which they called APSO.[12]

LHASA automatically displayed retro-reactions that might be suitable for a query structure. APSO did not do this. Instead, having entered a structure the student was first invited to name functional groups and significant features in it, such as bridgehead positions identified by the program. The program responded by telling the student whether he/she was right or wrong. The student then selected a functional group for further processing. The program presented all the reactions known to it that could generate the functional group, and the student was invited to decide which reaction was most appropriate. The program checked the feasibility of the choice, taking into account the rest of the query structure, and told the student whether it was a good one. If the choice was not a good one, the program explained why. If it was a good choice, the program entered into a dialogue with the student about reaction conditions and gave further feedback on the student's choices.

There are computer applications for teaching about things such as the periodic table and how to name chemicals, and Joyce Brockwell and John Werner have developed a system for teaching about organic synthesis,[13] but, apart from those applications, rather limited attention still seems to have been given to knowledge-based software for teaching about chemical synthesis, toxicology and related topics, such as metabolism and biodegradation.

References

1. L. C. Maillard, *Genèse de Matières Protéiques et des Matières Humiques: Action de la Glycérine et des Sucres sur les Acides α-Aminés*, Massonet et Cie., Paris, 1913.
2. K. D. Stewart, M. Shiroda and C. A. James, Drug Guru: a Computer Software Program for Drug Design Using Medicinal Chemistry Rules, *Bioorg. Med. Chem.*, 2006, **14**, 7011–7022.
3. M. Ott, *Developing an Expert Computer System to Predict Degradation Pathways*, presented at Informa Life Sciences 3rd Annual Conference on Forced Degradation Studies, Brussels, 27–28th January, 2009.
4. Zeneth is being developed by Lhasa Limited, 22-23 Blenheim Terrace, Woodhouse Lane, Leeds LS2 9HD, UK, and the description of Zeneth in this chapter is based on personal communications with Martin Ott at Lhasa Limited.
5. *Guidelines on the Limits of Genotoxic Impurities*, European Medicines Agency, London, 28th June 2008.
6. J. C. Dearden, Prediction of Environmental Toxicity and Fate Using Quantitative Structure-Activity Relationships (QSARs), *J. Braz. Chem. Soc.*, 2002, **13**, 754–762.
7. P. Reuschenbach, M. Silvani, M. Dammann, D. Warnecke and T. Knacker, ECOSAR Performance with a Large Test Set of Industrial Chemicals, *Chemosphere*, 2008, **71**, 1986–1995.
8. J. P. Hickey, A. J. Aldridge, D. R. May Passino and A. M. Frank, An Expert System for Prediction of Aquatic Toxicity of Contaminants, in *Expert Systems for Environmental Applications*, ed. J. M. Hushon, ACS Symposium Series No. 431, Am. Chem. Soc., Washington, DC, 1990, pp. 90–107.
9. P. Von der Ohe, R. Kühne, R. -U. Ebert, R. Altenburger, M. Liess and G. Schüürmann, Structural Alerts – a New Classification Model to Discriminate Excess Toxicity from Narcotic Effect Levels of Organic Compounds in the Acute Daphnid Assay, *Chem. Res. Toxicol.*, 2005, **18**, 536–555.
10. R. Kühne, F. Kleint, R. -U. Ebert and G. Schüürmann, Calculation of Compound Properties Using Experimental Data from Sufficiently Similar Chemicals, in *Software Development in Chemistry 10*, ed. J. Gasteiger, Gesellschaft Deutscher Chemiker, Frankfurt, 1996, pp. 125–134.
11. Unpublished work in a collaboration between Lhasa Limited, 22–23 Blenheim Terrace, Woodhouse Lane, Leeds LS2 9HD, UK, and Helmholtz-Zentrum für Umweltforschung GmbH – UFZ, Permoserstraße 15, 04318 Leipzig, Germany.
12. R. D. Stolow and L. J. Joncas, Computer-Assisted Teaching of Organic Synthesis, *J. Chem. Ed.*, 1980, **57**, 868–873.
13. J. C. Brockwell and J. H Werner, *Beaker: Expert System for the Organic Chemistry Student Version 2.1 DOS*, Wadsworth Publishing Company, Belmont CA, USA, 1995.

CHAPTER 18

Evaluation and Validation of Knowledge-Based Systems

There are difficulties with measuring and reporting on the effectiveness of knowledge-based prediction systems. The measures normally used for evaluating statistical models may not be appropriate for knowledge-based systems, but they are nevertheless the ones usually chosen. They are "sensitivity", "specificity" and "concordance"; the proportion of active compounds correctly predicted to be active; the proportion of inactive compounds correctly predicted to be inactive; and the proportion of the full set of compounds predicted correctly to be active or inactive, respectively:

let 'a' be the number of compounds predicted to be active;

let 'b' be the number of compounds that actually are active;

let 'c' be the number of compounds predicted to be inactive;

let 'd' be the number of compounds that actually are inactive.

Total number of compounds in the set $= b + d$

Sensitivity $= a/b$

Specificity $= c/d$

Concordance $= (a + c)/(b + d)$

You cannot judge performance on sensitivity alone, of course, which is why specificity and concordance also matter. To take the extremes, if a model always predicts everything to be active, it will achieve 100% sensitivity, and if it always predicts everything to be inactive, it will achieve 100% specificity. Concordance on its own does not give you any guidance on what bias there may be in the model to err, so to speak, on the side of caution or laxity.

The first weakness in this approach to assessing performance, applies to the evaluation of statistical models as well as knowledge-based systems. As Paracelsus is much quoted as having said in the middle of the fifteenth century,

RSC Theoretical and Computational Chemistry Series No. 1
Knowledge-based Expert Systems in Chemistry: Not Counting on Computers
By Philip Judson
© Philip Judson 2009
Published by the Royal Society of Chemistry, www.rsc.org

"toxicity is in the dose" – though probably not word for word, especially as his first language was not English. There is no absolute distinction between toxic and non-toxic chemicals. We impose thresholds convenient to us to define what amounts to being toxic in different contexts. The globally harmonised system for classification and labelling requires a chemical to be labelled as toxic, if its rat oral LD_{50} value (the dose leading to the death of half the rats in a study) is less than, or equal to, $300\,mg\,kg^{-1}$. You may or may not wish to take issue with the justification for such experiments, and the regulations provide alternatives, but the example illustrates the problem with trying to define toxicity. A compound with $LD_{50} = 299\,mg\,kg^{-1}$ is not classed as toxic, while one with $LD_{50} = 300\,mg\,kg^{-1}$ is. So for the purposes of evaluation of a model, performance against this threshold is what matters.

The problem of what to do about borderline chemicals when setting up training sets for models has been much discussed, and various solutions are applied, of which the most drastic, but probably also the most satisfactory, is simply to leave them out of the studies. The problem of what to do about borderline *predictions* when evaluating a model – or comparing it with other models – is less discussed and less satisfactorily resolved. Most usually, if a model predicts a numerical value, a threshold is decided upon and predictions are classed as positive or negative according to which side of the threshold they fall. Those that fall close to the threshold, but on the wrong side of it, are classed as failures. This is not satisfactory, because the model is at least correct in predicting that the compounds will have borderline toxicities but, so to speak, gets no credit for it. To add to the confusion, models that predict potency normally express a measure of confidence in the values they report in the form "$280 \pm 25\,mg\,kg^{-1}$". How do you incorporate the significance of that prediction into a single, overall measure of performance for the model it came from, if the threshold for activity is $300\,mg\,kg^{-1}$?

The problem is arguably worse for a knowledge-based system – certainly it is no better. In order to calculate values for sensitivity, specificity or concordance, a researcher must decide, for example, whether to treat all predictions from Derek for Windows that toxicity is at least plausible as positive predictions, to treat predictions that toxicity is at least probable as positive, or to treat only predictions that toxicity is certain as positive. In the last case, the evaluation of the model would be pretty meaningless, since Derek for Windows only regards toxicity as certain if there are positive results from actual laboratory tests on the query compound itself, and people doing evaluations normally choose plausible or probable as their threshold. The evaluations are subjective in this respect because whether plausible or probable is chosen usually depends on which one gives the better overall predictive performance. But how else is the choice to be made? There would be no sense in deliberately choosing the threshold that had the worst performance, and making an arbitrary choice would confuse things while achieving nothing. What do you do about "certain" predictions? In terms of statistical prediction models, it looks like cheating to include them, since they are not predictions but statements about observations (I would not be so rash as to say "statements of facts" in relation to biological data). On the other hand, if you are trying to assess the practical usefulness of a system it does not make much sense to exclude them.

Do not place too much confidence in every evaluation you look at. I am aware of some studies, happily not published, in which the triggering of an alert in Derek for Windows was interpreted as a positive prediction regardless of the likelihood associated with the prediction. Since rules in Derek for Windows can assign a likelihood of activity of "doubted", "improbable" or "impossible", all of which are predictions of *inactivity*, there was a fundamental flaw in the studies. In the case of Derek for Windows, the flaw is obvious once spelt out. Similar errors of interpretation may be less obvious for some other systems or models, and you need to look at evaluations and comparisons with circumspection.

The second weakness in evaluations, concerns an assumption universally made about knowledge-based programs by researchers who are used to assessing the performance of statistical models, namely that when a program does not predict activity it amounts to the same thing as predicting inactivity. That is not true. Derek for Windows, for example, most usually states either a positive likelihood that a compound will be active or that there is "nothing to report". In the latter case, the program is stating that the situation is open: it is not even stating that there is conflicting evidence; it has found no evidence for or against activity. It predicts that a compound will be inactive if, and only if, there are arguments directly for the case that it will be inactive, and then it is explicit about it.

The assumption that "open" and "negative" mean the same thing looks like a serious error, and in some circumstances it is. However, it may be all right in practice for predictions of an end-point such as mutagenicity by Derek for Windows, because mutagenicity is very well covered in the knowledge base. The chances of entering a query structure containing an alert that Derek for Windows does not know about are low. So if Derek for Windows fails to find a reason to suspect the compound of being active it might be all right to assume that the compound is more likely inactive. That is not the case with all end-points. Hepatotoxicity (liver toxicity), for example, is complicated, not well understood, and has only limited coverage in Derek for Windows. It would not be wise to assume that a compound that failed to trigger an alert in Derek for Windows would not damage the liver.

Neither is it correct to treat "nothing to report" as being the same as "unable to process your query". To put it in more colloquial terms, "nothing to report" in Derek for Windows means "I have looked for evidence that your compound might be expected either to be active or to be inactive but found nothing".

That leads us to weakness number three in evaluations – and especially comparisons between programs. A program may be unable to handle some kinds of chemical structure: some programs have limits on the total number of atoms in a structure that can be processed; others only deal with elements usually found in organic compounds – they might not accept structures containing transition metals, for example. Oncologic is limited to structures built from a pre-defined set of substructures and so are some other systems.

Statistically-based models usually have a defined applicability domain – the chemical space within which they were generated – and programs warn users if a query structure is outside the domain. Herein lies potential for confusion, or even – dare we say it? – for the convenient selection of performance statistics to suit marketing purposes. If your question is "how precise is the program at making predictions within its area of coverage?", you calculate sensitivity, specificity and concordance on the basis of the predictions made by the program. Those predictions, obviously, are limited to the structures which the program can handle. If your question is "how useful is the program for a particular end-point?", the measure based on its performance within its area of coverage is misleading. Suppose one model has a concordance of 90% within its area of coverage (*i.e.* it makes correct predictions for 90% of the structures that it can process), but it is a specialised model and only able to process 50% of the structures in a test set. Suppose that a second model has a concordance of 60%, but is able to process 95% of the structures in the test set. The first model successfully predicts for 45% of the compounds. The second model successfully predicts for 57% of the compounds. Which model is the more precise? Which is the more useful? Be careful about the question you ask. Be careful about the answers you get.

And now we come to the concept of an applicability domain (alternatively called a "prediction space"). You do not need to go into statistics or even to understand much about toxicology to realise that if you build a model to predict corrosivity using a training set of seventeen alkanesulfonic acids the model will be unlikely to make reliable predictions for compounds such as benzene, ethanol, or sodium hydroxide. In this case, chemical intuition tells us that the applicability domain for our model is a very narrow one – namely, alkanesulfonic acids. There is room for some uncertainty even so. Suppose the seventeen members of the training set cover chain lengths from three to twenty-six fairly evenly spread, but happen not to include octanesulfonic acid (chain length eight). Presumably you will be comfortable about using the model to make a prediction for octanesulfonic acid. What about methanesulfonic acid and triacontanesulfonic acid (chain length thirty), which are below the lower and above the upper ends of the range covered in the training set, respectively? If you are willing to take a chance on those, are you willing to extend the coverage to benzenesulfonic acid?

It is common to base the applicability domain for a QSAR model on the properties that feature in its equation. The most important property for a lot of QSAR models is log P (log K_{ow}), the octanol–water partition coefficient of the chemical of interest. If log P for a query compound lies within the range covered by the compounds that were used to train the model, the compound is considered to be within the applicability domain of the model. More sophisticated approaches take into account how dense the coverage in the training set was of log P values close to the value for the query, and/or how well the model performed in that region against a test set. It makes sense to define an applicability domain in this way. It is hard to think of anything likely to influence the reliability of a model more than the properties that drive it.

Finding a way to define the applicability domain for a model that is based on chemical substructures or structure-based descriptors such as linear fragments, atom pairs, or circular fingerprints, and deciding which are the better ones to use is more perplexing. The Multicase software, for example, analyses training sets in terms of automatically generated, linear structural fragments (see Chapter 11.3); the system developed by Gerrit Schüürmann's team in Leipzig (see Chapter 17.7) uses atom-centred fragments. It may seem straightforward to define the applicability domain in terms of fragments of these kinds – if all the fragments present in the query structure are included in the training set that was used for the model, then the query is within the applicability domain. But it is not so simple. If you allow fragments to be too precise and you generate all those that are possible, pretty well every structure will contain something unique. Hence nearly every query will be classed as outside the applicability domain. In practice, a degree of vagueness has to be allowed in the descriptors used to define the applicability domain and/or queries have to be allowed to contain a small number of unrecognised descriptors.

When it comes to deciding on a meaningful applicability domain for an alert or rule in a knowledge-based system, new questions come up. The better rules and alerts in a well-thought-out knowledge base will have a mechanistic basis. Some of them may be based on a limited set of examples, but with sufficient evidence about the mechanism for a human expert to make valid extrapolations. A human expert considering a query structure would have no clear preconceptions about an applicability domain. Being aware of the mechanism and the evidence that led to the writing of the alert he/she would consider the structural features of the query in that context in order to judge the reliability of making a prediction. So if you come up with a way of deriving a generic applicability domain based on generating descriptors from the structures that informed the writer of an alert, will it be satisfactory? Claire Ellison has used fragments generated by deleting lines from the connection tables in Mofiles for the structures in a training set[1] as a first step in exploring some of the issues.

Why are knowledge-based systems evaluated and compared on the basis of seemingly-inappropriate statistical performance measures, and what does it matter whether applicability domains can be defined for them?

Firstly, for toxicity prediction systems to be useful to industry they must be acceptable to the regulatory authorities. The European Centre for the Validation of Alternative Methods was set up to make recommendations on the acceptability of *in vitro* tests in place of *in vivo* tests, but has also taken a key role in encouraging the evaluation of *in silico* alternatives (*i.e.* computer methods).[2–4] An international workshop recommended criteria that QSAR methods should meet in order to be acceptable for regulatory purposes[5] which were adapted and adopted by the OECD[6] – one change being to refer to "(Q)SAR", making it explicit that the same principles should apply to non-quantitative models. Legislation and regulation now promote the use of *in silico* methods wherever possible but they need to meet the criteria if they are to be used. That means, among other things, that they need to give information to the user about whether a query falls inside the prediction domain for the model.

Secondly, no competent scientist should place trust in a research-based model unless all the necessary evidence is available to allow a proper assessment of the quality of the model. Actually, this second reason is ultimately the only one, since the requirements for acceptance by the regulators originate from their quest for scientific rigour. In the absence of any respectable scientific alternatives, the criteria developed for assessing the quality of statistical models in general are the ones currently being used for toxicity prediction models, both statistical and otherwise.

The OECD guidelines do not stipulate what techniques are to be used to measure the reliability of prediction models or how applicability domains are to be defined, but they do require that reliability is meaningfully demonstrated and that users have guidance on the scope of models. They state that to be acceptable for regulatory purposes a model should have:

a defined end-point;

an unambiguous algorithm;

a defined domain of applicability;

appropriate measures of goodness-of-fit, robustness and predictivity;

a mechanistic interpretation, if possible.

By "defined end-point" is meant one that is understandable and reproducible. For example, if your end-point is "mutagenicity" do you mean "mutagenic *in vivo*" and, if so, in what species and under what circumstances, or do you mean "giving a positive result in the Ames test", in which case, which version of the Ames test (there are many variations using different strains of bacteria and procedures)? The term "unambiguous algorithm" means that it must be fully clear to the user how predictions are generated – a condition not met by "black box" models, where you put data in, get an answer out, but have no idea why the answer is the way it is. Knowledge-based systems are well placed to satisfy both of these requirements: predicting for a defined end-point is straightforward and simply the responsibility of the developer of a model; models in knowledge-based systems inherently "know" how they make their predictions, and so these systems are best placed of all to meet the requirement for an unambiguous algorithm, as long as the thinking on the part of the rule writer is fully communicated to the end user.

Performance measures like sensitivity, specificity and concordance look like precise measures of predictivity, being numerical, but they are not. The way they are calculated is precise, but the input to them involves subjective decisions even when you are using quantitative models. Douglas Bristol, a scientist at the National Institutes of Environmental Health Sciences, drew my attention to the following comments from a paper he had been reading:

"Verification and validation of numerical models of natural systems is impossible. This is because natural systems are never closed and because results

> *are always non-unique. Models can be confirmed by demonstration of agreement between observation and prediction, but confirmation is inherently partial... The primary value of models is heuristic."*[7]

The final sentence in that quotation points to better ways for a user to assess how models in a knowledge-based system are likely to perform for a particular query. For example, a system that classifies chemicals according to their likely potency might provide information to the user on the chemicals from a test set that it placed in each category, together with the potencies of the chemicals that were found experimentally. Given features in the program for viewing the data in different ways, the user could explore the information in the context of the query of interest and learn more about the likely reliability of a prediction than he/she would get from the values for sensitivity, specificity and concordance, even if those values were trustworthy.

Meeting the last OECD guideline, providing a mechanistic interpretation, is the particular strength of knowledge-based models. Many rules are written by gaining an understanding of a particular mechanism of action, and when writers start out by writing rules to capture simply what has been observed, they routinely go on to seek out at least putative mechanistic explanations which they incorporate into notes associated with the rules.

The developers of knowledge-based systems aim to be fully compliant with the guidelines.[8] Some guidelines are already met, but that may not have been stated explicitly in the past. Defining the domain of applicability is not as clear cut for any model as is sometimes suggested, and it is potentially tricky for some kinds of rules in knowledge-based systems. By precedent, measures of goodness-of-fit, robustness and predictivity are taken to be numerical, and usually to be those used in statistics. There are likely to be difficulties in gaining acceptance for measures more appropriate to knowledge-based models. Knowledge-based systems should be able to meet the OECD guidelines well, but there is work to be done and, as you will deduce from what is written earlier in this chapter, the ways in which some of the guidelines can best be met are likely to be hot topics of debate.

References

1. C. M. Ellison, S. J. Enoch, M. T. D. Cronin, J. C. Madden and P. Judson, A Structural Fragment Based Approach to Define the Applicability Domain of Knowledge Based Predictive Toxicology Expert Systems, *Alternat. Lab. Animals*, in press.
2. J. C. Dearden, M. D. Barratt, R. Benigni, D. W. Bristol, R. D. Combes, M. T. D. Cronin, P. N. Judson, M. P. Payne, A. M. Richard, M. Tichy, A. P. Worth and J. J. Yourick, The Development and Validation of Expert Systems for Predicting Toxicity. The Report and Recommendations of an ECVAM/ECB Workshop (ECVAM workshop 24), *Alternat. Lab. Animals*, 1997, **25**, 223–252.

3. A. P. Worth and M. T. D. Cronin, Report of the Workshop on the Validation of QSARs and Other Computational Prediction Models, *Alternat. Lab. Animals*, 2004, **32**, 703–706.
4. S. Coecke, H. Ahr, B. J. Blaauboer, S. Bremer, S. Casati, J. Castell, R. Combes, R. Corvi, C. L. Crespi, M. L. Cunningham, G. Elaut, B. Eletti, A. Freidig, A. Gennari, J.-F. Ghersi-Egea, A. Guillouzo, T. Hartung, P. Hoet, M. Ingelman-Sundberg, S. Munn, W. Janssens, B. Ladstetter, D. Leahy, A. Long, A. Meneguz, M. Monshouwer, S. Morath, F. Nagelkerke, O. Pelkonen, J. Ponti, P. Prieto, L. Richert, E. Sabbioni, B. Schaack, W. Steiling, E. Testai, J.-A. Vericat and A. Worth, Metabolism: a Bottleneck in *In Vitro* Toxicological Test Development. The Report and Recommendations of ECVAM Workshop 54, *Alternat. Lab. Animals*, 2006, **34**, 1–36.
5. J. S. Jaworska, M. H. I. Comber, C. Auer and K. J. Van Leeuwen, Summary of a Workshop on Regulatory Acceptance of (Q)SARs for Human Health and Environmental Endpoints, *Environ. Health Perspect.*, 2003, **111**, 1358–1360.
6. *OECD Principles for the Validation, for Regulatory Purposes, of (Quantitative) Structure–Activity Relationship Models*, published 19th December 2006 and available for download free of charge from the OECD website, http://www.oecd.org/. The address of the OECD is 2, rue André Pascal, F-75775 Paris, Cedex 16, France.
7. N. Oreskes, K. Schrader-Frechette and K. Belitz, Verification, Validation, and Confirmation of Numerical Models in the Earth Sciences, *Science*, 1994, **263**, 641–646.
8. K. Langton, *Conforming to the OECD Principles for (Q)SAR Validation*. A paper presented at the 14th Congress on Alternative to Animal Testing, Linz, 2007. Available from the Lhasa Limited website, http://www. lhasalimited.org/

CHAPTER 19

Combining Predictions

19.1 Existing Approaches to Combining Toxicity Predictions

There is a lot of interest in comparing the output from different kinds of programs or models in the hope of improving the overall success of predictions. Frequently called "consensus modelling" it sometimes does depend on looking for the areas of agreement between applications as "consensus" would imply, but more usually it is a weight-of-evidence approach. For example, when Anita White and colleagues compared TopKat, Derek for Windows, and Multicase side by side, they looked at the use of voting systems to combine the outputs in order to improve the overall reliability of predictions.[1] Makoto Hayashi and colleagues have described similar work at the National Institute of Health Sciences (NIHS) in Japan.[2,3] They compared predictions of mutagenicity (as evidenced by positive results in the Ames test and chromosome aberration tests) from Derek for Windows, Multicase, and ADMEWorks.[4] They found that assigning an overall positive or overall negative prediction to the cases where all three programs agreed greatly improved the reliability of the predictions (but, of course, at the cost of having a lot of cases for which no prediction could be made – the cases where at least one program disagreed with the other two). In their studies, they found that accepting predictions where at least two of the programs were able to predict and were in agreement, gave a significant improvement in overall reliability compared with using an individual program, and they concluded that, although the reliability was not as high as when all three programs agreed, accepting a majority vote was the better compromise for practical purposes. It has been said in many ways, by many people, that you cannot expect to get good predictions by combining a lot of bad ones, and there are other issues, although the researchers at the NIHS were aware of them and tried to avoid the pitfalls.

RSC Theoretical and Computational Chemistry Series No. 1
Knowledge-based Expert Systems in Chemistry: Not Counting on Computers
By Philip Judson
© Philip Judson 2009
Published by the Royal Society of Chemistry, www.rsc.org

A dangerous assumption, discussed more fully in Chapter 18, is the one that when a knowledge-based program does not predict activity it amounts to the same thing as predicting inactivity. As explained more fully in Chapter 18, that is not true. However, as also explained in Chapter 18, the assumption is approximately true, and may be all right, for predictions about end-points that are covered well by a program, which was the case in the work on prediction of mutagenicity by Anita White, Makoto Hayashi, and their respective colleagues.

There is host of issues to do with the independence of the predictions you are combining. To take an obvious example, suppose that you use a simple voting scheme with three prediction systems: if any two of the systems agree that a query will be toxic, or that it will not, you accept that prediction. What if two of the systems use the same, or a very similar, modelling method developed from the same data set? Of course they will agree (assuming they are both more or less bug-free). They will both be right and wrong in concert, which is not good news, and the third system will have as much relevance as would minor parties in a parliament where the government and the main opposition agreed about everything.

If you want to bring together systems that use different modelling methods you cannot include "black box" solutions where the vendor keeps the methods secret. Fortunately, there are no commercial toxicity prediction products in that category – you know the methods used, even if you do not know the details of the implementations in some cases. It is less easy to be sure about the data sets used for training models. There is a much-lamented dearth of toxicological data in the public domain suitable for (Q)SAR modelling work. So in the absence of evidence to the contrary, it may be best to assume that models built from published data sets will have used largely the same data. A further problem is that the vendors of some models do not disclose the structures that were used to develop them, or, more prevalently, a significant proportion of structures is not disclosed because model builders have been given access to confidential structures by regulatory bodies or companies to supplement the public data. Even if two models were built using confidential data from different sources, there is no way of knowing how many structures might be in common between them.

It is not necessary for models and their training sets both to be different, for them to be useful in making combined predictions. You may be able to take advantage of the different strengths and weaknesses of modelling methods even though they have been built from the same data. Toxicology data sets are notoriously difficult to work with. Almost all public data sets, and probably most proprietary ones, are heavily biased either towards active compounds or inactive ones. A database of carcinogenic compounds, for example, by definition will be populated with active carcinogens. A database of compounds submitted to a regulatory body as part of a registration package for a pharmaceutical or agrochemical application will be biased in favour of non-genotoxic compounds, because those found to be highly active will have been filtered out of product development long before the stage at which a company would submit a registration package. Models built by the same method, but from different data sets, may help to even out bias in the data sets.

Bias is only one of the problems. There are so many toxicological experiments and so many variants of each, that two models, based on different training sets and apparently predicting the same end-point using the same criteria, may actually be predicting subtly different things. In a typical toxicological database a substantial percentage of data will be misleading – not because researchers or database staff have been careless, but because biological experiments perform so variably. The first step before setting about building a model is to go through the data to be used for training and testing, and to take out suspect entries if they can be identified. Even so algorithms for generating models need to be able to cope with high noise levels in the data, and once models have been built they need careful assessment to make sure that they have not been influenced by systematic errors in the data. If you want to combine predictions from different models it is important to establish with what diligence the models were created. It may not be necessary or desirable to exclude models, but there may be reasons to have more confidence in predictions from some than from others.

Finally, as Makoto Hayashi's team pointed out,[3] you need to think about the criteria you set for classifying a compound as toxic on the basis of multiple predictions. If, for example, you decide that you will classify a compound as active if three models all predict it to be so, and one of the models can handle only a narrow range of structures, your overall prediction system will be similarly constrained. In cases where that model could not process your query, would you not want to know about the predictions from the models that could handle it rather than to get nothing?

A popular way of bringing together multiple predictions while avoiding issues about combining them is simply to present them all to the end user, but in a convenient way. At least half a dozen companies have in-house applications providing this sort of service.[5] The predictions from different systems or models are presented in tables, typically implemented as spreadsheets in Microsoft Excel because it is available to most PC users. Several of the systems I have seen use simple colour coding to draw the attention of the user to the more interesting cases. Not surprisingly, the popular option is to colour the background of cells in the spreadsheet red where high toxicity is known or predicted, yellow where there is moderate reason for concern, and green if there is no clear reason for concern. So if all the cells across the spreadsheet are red for a compound, it looks like one to get worried about. If there is a mixture of colours across the page, the disagreement between different assays and/or prediction models is evident, and a more detailed investigation of the data is called for.

Using background coding in this way does not abrogate the more specific information that each model offers. For example, whether you colour cells in the column for predictions from a model red, if the value it estimated was greater than x, and yellow, if it was between x and y, or red in both cases, the values themselves can still be displayed in the boxes – information is not lost. The spreadsheet approach brings together assay data and predictions from different sources, instead of comparing results automatically in order to produce a single output. The end user is empowered to form his/her own views on the merits of the different predictions and to come to a reasoned conclusion.

19.2 The OECD (Q)SAR Toolbox

The Organisation for Economic Co-operation and Development (OECD)[6] is running a project, funded by the European Union and in collaboration with regulatory bodies in Europe, Japan, and America and the International QSAR Foundation,[7] to provide a (Q)SAR toolbox suitable for use by the regulators. A prototype, developed for the project by the Laboratory of Mathematical Chemistry[8] under the leadership of Ovanes Mekenyan can be downloaded free of charge.[9] The toolbox is not intended to make predictions about toxicity, but to help a user who wants to form a view about potential toxicity to find information and/or to develop suitable prediction models. It is nevertheless pertinent to this chapter, because being a way of bringing together information from multiple sources, it needs to address many of the issues relevant to bringing together multiple predictions.

The toolbox is designed around the workflow a regulator would typically follow. The first step is to identify the chemical of interest, and the toolbox allows you to enter a structure or a SMILES code, or to search in databases to which it has access, using a chemical name, or a code such as a CAS number or a European chemical inventory code. The system then puts together a profile for the chemical, drawing information from its databases. The databases include inventories of the contents of other, independent databases such as the HPVC[10] and TSCA[11] databases, and so the profile advises on whether the query compound is included in them as well (but does not directly provide the data contained in them). The profile also covers broad classifications of the chemical, such as whether it is defined as organic or inorganic (relevant to OECD regulations about labelling for transport, for example), whether it is a single compound or a mixture, whether it is a polymer, *etc.* In addition, it lists how the chemical fits into systems for classifying chemicals that are used for some regulatory purposes, for example whether it is a nitro compound, an aldehyde, *etc.* If hazard and/or risk assessments already exist, they are included, and so is information based on structural alerts in the chemical. The purpose of alerts is not to predict toxicity, but to classify the chemical with others with which it might share toxicological properties, and which might therefore be useful reference points for making judgements about the chemical.

The next step in the workflow is to collect existing data about the chemical, if there are any. The user will not often find such information in practice, since the most likely reason for doing the search is that the chemical is new to the regulators, but there will be cases where another regulator, maybe on the far side of the globe, holds data that the user might not discover without the help of the toolbox. Perhaps sometimes the necessary data will come to light and the search will be over at this stage. More usually it will be necessary to move to the next step – category definition. The user can categorise the query according to several schemes. The purpose of this is to be able to seek out those chemicals which appear to be the best analogues from which to deduce predictions about the query. Looking for similarity in terms of, say, the presence of common substructural features is not enough by itself. In the opinion of the designers of the toolbox (and I agree), the features in common that matter are ones that are associated with specific mechanisms of toxicological action. Substructures are themselves grouped into categories, such as those associated with protein binding. In this context other properties in common may be just as important as the presence of particular substructures – *e.g.* having similar solubility or partition properties, or similar acidity or basicity. Chemicals that would be expected to differ significantly from the query, because of likely metabolism or mechanisms of toxicity that are not appropriate to the query, are eliminated.

The chemicals that have been identified as relevant, and the data associated with them, are presented in a spreadsheet.

The user is now ready to set about data gap filling, *i.e.* to estimate data values for the query where there are gaps in the spreadsheet. The simplest approach offered is read-across – *i.e.* interpolation between values of the property of interest for a series of related compounds that follow a consistent trend – but the standard statistical methods needed to put together a new QSAR model are provided in the toolbox.

Finally, the user is ready to issue a report, and the system supports the process, ensuring that everything the user has done is documented. Subsequently, therefore, any regulator should be able to repeat the process and come to the same conclusion.

In terms of relevance to this chapter, the significant point about the OECD (Q)SAR Toolbox is that it provides a "one-stop shop" for bringing together data and predictions from diverse sources and working with them in a standardised environment.

19.3 Combining Predictions about Modes of Action that are Largely Independent

Chapter 17.7 describes work done by Gerrit Schüürmann's team on narcosis and excess toxicity in fish and daphnia. It provides an example where predictions from different models can be reliably combined in a single package. Predictions of narcosis are based primarily on physical properties and the mechanism of action is believed to be through disruption of cell membranes. Excess toxicity arises from specific, more catastrophic interference with cell processes such as the damage caused by strong acids, or through mechanisms involving direct interaction with active or allosteric sites in proteins. Prediction of excess toxicity is triggered by the presence of atom-centred fragments or sub-structural alerts in Schüürmann's system. So questions do not arise about how to combine two models that both predict the same end-point by the same mode of action, or about which prediction should dominate: the modes of action *are* different and by definition the effects of excess toxicity, if it occurs, will eclipse those of narcosis. If the program calculates the acute toxicity arising from narcosis and simply augments it, or over-rules it, with predictions of excess toxicity if they arise, the resultant output will be satisfactory. At some future point, of course, the need might arise to make predictions about the relative importance, or interactions between, different mechanisms of excess toxicity and then all of the issues discussed in Chapter 19.1 will come into play, but, for the present, dealing with the relationships between narcosis and excess toxicity is an advance on what systems have previously done.

19.4   Combining Metabolism Predictions – the NoMiracle Project

The NoMiracle[12] project is a typical Framework 6 project funded by the European Union. It involves thirty-eight organisations located from Denmark to Spain, England to Bulgaria, and enough countries in between for you to walk from partner to partner, without ever leaving home territory for the project. Its core scientific objectives are to improve understanding of the distribution of and interactions between different pollutants in the environment and the consequences for environmental and human health. The project is divided into four pillars, and within one of them was a very small sub-project on improving the prediction of environmental degradation led by Ovanes Mekenyan in Bulgaria[8] in collaboration with Lhasa Limited in England.[13] Although at the time of writing of this book, the five year NoMiracle project is still running, the sub-project on predicting environmental degradation was a three year one and has already ended. The work was described at a conference in 2008[14] and a more detailed report should become available from the NoMiracle web site.[12]

The sub-project investigated the combined reporting of data and predictions from different sources using computer reasoning methods. For the purposes of the study, two prediction systems and two databases of biodegradation reactions were used – CATABOL, Mepps, a database that is supplied with CATABOL, and data from the University of Minnesota Biocatalysis/Biodegradation Database (see Chapters 16.2 and 16.3). This set of programs and databases would not be ideal for a system intended for serious use because there is known to be significant overlap between them, but they were sufficient for the purposes of the project, which were to investigate the issues with trying to compare different systems, developing a scheme for assessing confidence in predictions, and producing reports that drew together the resultant information.

It quickly became apparent that comparing the biodegradation trees generated by different systems would be difficult. Some single-step reactions in one system were treated as two-step reactions in another. Very often different systems showed what were, for practical purposes, equivalent reaction sequences, but with reaction steps in a different order. Some differences might be really trivial, others more significant. The potential users of a biodegradation prediction system are more interested in what compounds might be formed than in the reaction sequences leading to them. So Mira, the application developed in the project, compares the sets of products generated by the different systems, rather than the biodegradation trees.

The most significant feature of Mira that distinguishes it from other reasoning-based applications developed at Lhasa Limited is that two reasoning chains operate in parallel. One generates advice about likelihood and the other about confidence in the prediction. Likelihood is reported using the qualitative terms that Mepps uses, such as "LIKELY", "VERY LIKELY". Confidence is reported using the terms "LOW", "MODERATE", "HIGH" and "VERY HIGH". For example, if the query compound is covered in one of the databases and the degradant is found there, confidence is "VERY HIGH"; if a degradant features in the biodegradation tree of only one predictive system, confidence is "LOW".

The demonstration prototype developed in the project is a web-based system. The user submits a query structure as a SMILES string, processing runs in batch on the server machine and the results are returned to the user in the form of written reports by email. In one report, advice is given about potential ease of biodegradation and about potential biodegradants, in the form of simple statements such as:

It is VERY LIKELY that the compound can biodegrade.
We have MODERATE confidence in this prediction.

In a second report, all the potential biodegradants that have been identified are listed as SMILES strings in a table suitable for opening in a program such as Microsoft Excel. An overall prediction of the likelihood that each biodegradant will be found is given, generated from a consideration of the output from both Mepps and CATABOL by applying reasoning of the kind described in Chapter 13.2, together with an assessment of the confidence attached to the overall prediction. CATABOL reports numerical probabilities for the biodegradants it proposes, which are converted to qualitative terms for the purposes of comparison in the reasoning.

19.5 Combining Different Models and Predictions about Different Properties

People have expressed a need for years to be able to bring together all the elements of "ADMET" – Absorption, Distribution, Metabolism, Excretion, Toxicity – in order to make balanced judgements about the potential safety and efficacy of candidate drugs. For the most part, the comments I have heard at scientific meetings have not been explicit requests for "umbrella" software solutions – more usually, they have been of the kind "It's all very well having this model for interaction with enzyme X, but what about gut absorption?" "Your proposed structure might make a good pharmaceutical for intravenous administration but if you give it orally won't it be metabolised rapidly in the liver?" – but the implications are clear enough: software that deals with each of the elements in ADMET independently has its uses, but if we want to get the best out of computing we must pull them all together.

Most of the existing systems for predicting toxicity have taken steps in this direction in that their predictions are moderated by estimated octanol–water partition coefficients, which to some degree influence both distribution and excretion of a chemical in mammals, but it is only relatively recently that the idea of creating over-arching software to bring everything together has started to be taken seriously. To turn the output from several toxicity prediction systems into a single "yes"/"no" answer automatically, you have to make difficult compromises and your decisions require you to anticipate what will be important to the end user. Makota Hayashi and colleagues, for example,[2,3] show that changing the way you bring together predictions from three systems can have opposite effects on the reliability of positive and negative predictions. The popularity of presenting multiple predictions in spreadsheets described in Chapter 19.1 conveys a message: presumably, since it is done, users can develop an overall interpretation of multiple predictions by applying human reasoning. I suggest that the way to avoid the weaknesses of generating a "yes/no" answer while still coming up with some kind of overall guidance, is to abandon "yes/no" answers in favour of qualitative terms supported with reasoned arguments so that users can make their own judgements.

Whether you use words like "probable" and "plausible" as in Derek for Windows, "very likely" and "likely" as in Mepps, class codes like 1, 2A, 2B, 3, 4 as used by the International Agency for Research on Cancer (IARC), or some other scheme does not matter. For example, on the basis of the findings of Makota Hayashi *et al.*, if three systems agree that a compound will give a positive result in an Ames test, the combined prediction might be at level 1 in a classification scheme; if two systems agree it might be at level 2; if only one system is able to give an answer it might be at level 3. A similar pattern might apply to predictions of inactivity. Your model for combining predictions is not restricted to the needs either of end users whose priority is for positive predictions to be correct, or whose priority is to avoid missing even the suspicion that a chemical might be toxic. The full range of information is expressed and the user can view or interpret it according to need.

Work is in progress in various projects to develop umbrella systems that bring together the different components of ADMET and draw on predictions from multiple models. In such an umbrella system, the physical properties of a query structure will be estimated; its mammalian toxicity will be estimated using alerts and rules in a knowledge base, appropriate QSAR models, and relevant information in a database; its potential metabolites will be generated and their properties and toxicities estimated; bringing all of these together, the system will be better able to give guidance on the effects likely to be seen in practice than current systems are.[15] New ideas are coming forward on the use of rules of thumb in place of numerical models for assessing ADMET properties for pharmacologically interesting compounds.[16]

An umbrella system for predicting, say, environmental degradation might make calls to a similar range of models but would see them from a different view-point. For example, the prediction of toxicity, especially to microbes, would have implications for biodegradation since a dead microbe is not going to contribute to it. If you consider the applications of knowledge-based systems described in this book, you will see how all of them would benefit from the use of umbrella systems, and it will not take you much thought to come up with others outside the scope of the book. To implement to the limit all that this implies would be a massive task, and it is not likely to happen, but we can expect real advances.

Combining different prediction methods might make it possible to draw on their different strengths, but there are yet hazards on the path. The user interfaces of reasoning-based, QSAR-based and other statistical systems are currently too wrapped up in their own technologies. Reasoning-based systems assume that users are comfortable with theoretical notions about logic and reasoning, and that they will explore crucial supporting information without being directed to it. As I commented in Chapter 3, part of the undoing of some of the synthesis planning systems was that users were expected to understand the theories behind them and the accompanying jargon. Many systems built around statistical analysis are worse than the reasoning-based systems when it comes to communicating with non-expert users. They require users to understand the uncertainties of model building and the pitfalls associated with measures of confidence such as r^2, q^2, and concordance. It is depressingly common to find computer systems delivering estimates of properties with extraordinary precision, even though they are actually very approximate, for no reason other than that the internal calculation is done with decimal variables and no-one has thought to use a format more appropriate for output. At least one database system reports the rat oral LD_{50} value for chloroform (trichloromethane) to six decimal places. If you look around you might find a system that uses "0" and "1" as shorthand for "negative" and "positive" but actually presents them on screen in scientific decimal notation, as "0.0000E0.00" and "1.0000E0.00". There are technical reasons why it does so, but it is not among the best examples of friendly communication.

Sweeping everything under the carpet and delivering apparently firm conclusions is not the answer. The acceptance of systems that draw together all the threads in ADMET, and of reasoning-based systems in general, will rest on how well user interfaces are designed to communicate fully, effectively and appealingly with users.

References

1. A. C. White, R. A. Mueller, R. H Gallavan, S. Aaron and A. G. Wilson, A Multiple In Silico Program approach for the Prediction of Mutagenicity from Chemical Structure, *Mutat. Res.*, 2003, **539**, 77–89.

2. M. Hayashi, E. Kamata, A. Hirose, M. Takahashi, T. Morita and M. Ema,
    In Silico Assessment of Chemical Mutagenesis in Comparison with Results
    of Salmonella Microsome Assay on 909 Chemicals, *Mutat. Res.*, 2005, **588**,
    129–135.
3. A. Hirose, presented at the New Horizons in Toxicity Prediction Sympo-
    sium, University of Cambridge, UK, 2008.
4. ADMEWorks, developed by Fujitsu Limited and its subsidiaries, is
    available from Fujitsu Management Services of America Inc., 1250E,
    Arques Avenue, Sunnyvale, California, USA.
5. Unpublished discussions with toxicology department staff in the research
    divisions of several leading pharmaceutical companies.
6. OECD, 2 rue André Pascal, F-75775 Paris, Cedex 16, France. http://
    www.oecd.org/.
7. International QSAR Foundation, 1501 W Knife River Road, Two Harbors,
    Minnesota 55616. http://www.qsari.org/.
8. Laboratory of Mathematical Chemistry, University "Prof. Assen Zlatarov",
    1 Yakimov Street, Bourgas, 8010 Bulgaria.
9. Information about the (Q)SAR Toolbox Project is provided on the OECD
    website: http://www.oecd.org/document/23/0,3343,en_2649_34379_
    33957015_1_1_1_1,00.html, from which the toolbox itself can also be
    downloaded.
10. The OECD High Production Volume Chemicals (HPVC) list is available in
    the form of a pdf file from http://www.oecd.org/dataoecd/55/38/
    33883530.pdf.
11. The Toxic Substances Control Act database (TSCA, usually pronounced
    "Tosca") can be accessed on-line *via* the Dialog web page: http://library.
    dialog.com/. Dialog's address is Dialog LLC, The Knowledge Center,
    11000 Regency Parkway, Suite 10, Cary, NC 27511. The database comes
    from the U.S. Environmental Protection Agency, TSCA Assistance Office,
    Office of Pesticides and Toxic Substances, 401 M Street S.W., MS-TS799,
    Washington, DC 20460, USA.
12. The web site for the NoMiracle project is: http://nomiracle.jrc.ec.
    europa.eu/.
13. Lhasa Limited, 22–23 Blenheim Terrace, Woodhouse Lane, Leeds LS2
    9HD, UK.
14. M. L. Patel, M. D. Hobbs, P. N. Judson, M. A. Ott, M. Ulyatt and J. D.
    Vessey, poster presented at 8th International Conference of Chemical
    Structures, Noordwijkerhout, The Netherlands, 2008.
15. Unpublished discussions.
16. M. P. Gleeson, Generation of a Set of Simple, Interpretable ADMET
    Rules of Thumb, *J. Med. Chem.*, 2008, **51**, 817–834.

CHAPTER 20

A Subjective View of the Future

As a society we make slow progress on resolving the difficulties of dealing with uncertainty. There is a belief that science is about facts; that science is about precision, and numbers are what we need. The consequent predilection for mathematical models and statistical measures of performance as the tools for every kind of prediction is dangerous. In putting our trust in numbers and our faith in the objectivity of science, we deceive ourselves. Albert Einstein is reported to have said "As far as the laws of mathematics refer to reality they are not certain; and as far as they are certain they do not refer to reality".[1] A study group at the Royal Society observed "The view that a separation can be maintained between 'objective' risk and 'subjective' or perceived risk has come under increasing attack, to the extent that it is no longer a mainstream proposition".[2] It is better to recognise and to deal with subjectivity than to pretend it is not there.

We accept reasoned judgements every day outside the confines of what we perceive as science. Doctors make diagnoses and recommend remedies, judges and juries decide the fates of fellow citizens, committees grant or reject applications about everything, from the opening and closing of railway lines, to the opening and closing times of night clubs. In conversation, many scientists express the opinion that qualitative methods for the prediction of toxicity can match or out-perform numerical ones; they place their greatest confidence not in formal models but in the collective views of human experts. Collective wisdom, or supposed wisdom – hardly more than a form of collective subjectivity – has led human societies to make some dire errors, but on balance it has been to our evolutionary benefit. The pursuit of true objectivity might not only be illusory, but a hindrance in the field of applied science.

Knowledge-based systems are in mainstream use for the prediction of chemical toxicity, and gaining acceptance for the prediction of metabolism, biodegradation and other chemical processes, but they do not yet have the scientific respectability accorded to statistical methods (at least by statisticians). Statistical

RSC Theoretical and Computational Chemistry Series No. 1
Knowledge-based Expert Systems in Chemistry: Not Counting on Computers
By Philip Judson

Published by the Royal Society of Chemistry, www.rsc.org

QSAR methods deliver quantitative answers, offering the appeal of apparent surety, while knowledge-based systems remind the user of uncomfortable uncertainties. Depending on your allegiances, you might want to rephrase that as "statistical QSAR methods deliver clear answers while knowledge-based systems are vague" but I hope that in the course of this book you have come to lean at least a little towards the different bias that my previous wording betrays.

The distinctions between the two approaches go deeper than just the difference between delivering quantitative and qualitative answers: the advocates of the two agree that uncertainty is inescapable, but the ways they believe it should be dealt with, are different. Statisticians are rightly frustrated by the general lack of understanding of even basic statistical concepts. The public are regaled daily by the press with prognostications screaming ignorance of simple statistics, but for every panic headline in a newspaper you can find a dozen papers in scientific journals that reveal their writers to have no better a grasp of the subject. The proponents of quantitative prediction believe that uncertainty can and should be quantified, too. It is accepted correct practice throughout science for values that have been measured, let alone predicted, to be qualified by error bounds (*e.g.* "99 ± 0.5"). The reliability of a prediction should be calculated and reported to the user (*e.g.* as r^2 and q^2 values). If the probabilities of events are predicted, those probabilities should be calculated according to the accepted formalisms of stochastic probability, the rules of chance. Simplistically expressed, they believe that the way to get better predictions is to develop better statistical models and to use them correctly.

Researchers who favour qualitative methods, believe that focussing on the machinery of statistics misses the point – namely that the problems lie with data and how to describe the world. Biological data are sparse, and will remain so given the size and complexity of the biosphere, and our coverage of chemistry is even thinner. In a recent talk, Chihae Yang[3] said, to paraphrase, that we know something, but not much, about ten million or so chemicals while the theoretical number of possible compounds is 10^{59}. We will not be filling the data gap for nearly 10^{59} compounds in the near future! Even if we can address the variability of biological data, we cannot hope to have representative datasets big enough for us to build generally applicable prediction models using statistical methods. On the other hand, using our generic understanding of chemistry and biochemistry, we are tolerably good at designing chemical syntheses, predicting chemical toxicity due to known mechanisms, and so on. Our best hope is to apply the human mind to what is known, in order to predict what is to come; we should put our faith in epistemic probability.

If you sit with the quantitative or qualitative supporters – let us call them the determinists and the genericists – and consider the weaknesses of the teams, you may latch onto failings on the other side without noticing that your side has them as well. Speaking for the genericists you accuse the determinists of basing predictions on apparent, but unexplained (and therefore suspect) correlations found through automated, statistical analyses; in reply, the determinists ask you for the evidence that some of the genericists' prediction models are based on real correlations, rather than on hunches. The genericists make predictions on the basis of what is already known, but, you complain from the terraces of the determinists, the mere application of existing knowledge cannot be described as prediction: the genericists reply that determinists say a model is valid only for a narrow domain of applicability, and you cannot get more confined to existing knowledge than that.

There is something bigger behind the debate about the relative merits of statistical QSAR and knowledge-based prediction. It is behind our concerns about how far we trust computer methods. More than that it underlies our concerns about how far we trust science. It is the question I introduced in the first paragraph of this chapter. Can science be truly objective or is subjectivity unavoidable – perhaps even beneficial? Reasoning-based computer systems give us the means to balance objectivity and subjectivity. It makes it possible for a computer to enter into dialogue with a human user in which both parties – if a computer can be called a party – express views about arguments and evidence, exchanging the kinds of contributions to a debate that start with "Yes, but . . . " or "What if . . . " and reach, we hope, consensus – a true alliance of subjectivity with objectivity.

If you have read this far, you must have some interest, at least, in where reasoning-based computer prediction is going. It at all depends on where you want to take it.

References

1. Attributed to Albert Einstein by D. Bristol, quoting from D. Brutlag, *Information Science for Molecular Biologists*, a tutorial notebook presented at the Intelligent Systems for Molecular Biology Conference, Stanford University, 1984, p. 52.
2. Royal Society study group, *Risk: Analysis, Perception and Management*, The Royal Society, London, 1992.
3. Comments made by C. Yang during a presentation at an internal seminar at the University of Sheffield, 5th December 2008.